ATMANIRBHAR BHARAT

Transforming Economy
Through Education,
Technology and Sustainability

ATMANIRBHAR BHARAT

Transforming Economy Through Education, Technology and Sustainability

AGRIEEM JASOLA
MADHU JASOLA

BLOOMSBURY
NEW DELHI • LONDON • OXFORD • NEW YORK • SYDNEY

BLOOMSBURY INDIA
Bloomsbury Publishing India Pvt. Ltd
Second Floor, LSC Building No. 4, DDA Complex,
Pocket C – 6 & 7, Vasant Kunj,
New Delhi 110070

BLOOMSBURY, BLOOMSBURY PRIME and the Diana logo are trademarks of Bloomsbury Publishing Plc

First published in India 2024
This edition published 2024

ISBN: 978-93-61315-48-0
2 4 6 8 10 9 7 5 3 1

Printed and bound in India by Replika Press Pvt. Ltd.

To find out more about our authors and books, visit www.bloomsbury.com and sign up for our newsletters

Contents

Preface

India stands at a transformative juncture in its history, poised to embark on a journey towards self-reliance and self sufficient nation. The vision of ATMANIRBHAR Bharat, or self-reliant India, is not just a slogan but a clarion call for a national movement aimed at harnessing the potential of every citizen, every sector and every resource in our country. This book, "ATMANIRBHAR BHARAT: Transforming Economy Through Education, Technology and Sustainability", brings together diverse perspectives, research and practical insights on education, technology and sustainability that chart the roadmap towards realizing this ambitious vision.

The concept of ATMANIRBHAR Bharat, introduced by Prime Minister Narendra Modi, encompasses five pillars: Economy, Infrastructure, System, Vibrant Demography and Demand. It calls for a robust economy driven by innovative and sustainable practices, world-class infrastructure that meets the needs of a growing population, technologically advanced and efficient system of governance, vibrant and youthful demography ready to drive change and a domestic demand that fuels local industries and enterprises. Each chapter in this book delves into these pillars, exploring how various sectors can contribute to and benefit from this transformative journey.

This book is a collective effort of distinguished scholars, industry leaders, policymakers and practitioners who have contributed their expertise and insights. Their contributions provide a comprehensive understanding of the challenges and opportunities that lie ahead. The chapters included herein offer a blend of theoretical frameworks, empirical studies and case analyses, making this book a valuable resource for students, researchers, policymakers and people interested in the socio-economic development of India.

The journey to ATMANIRBHAR Bharat is not without its challenges. As we navigate the complexities of a globalized world, the need for innovation, resilience and adaptability becomes paramount. This book emphasizes the importance of collaborative efforts, both within the country and with the global community, to achieve sustainable development and inclusive growth. It underscores the role of education, technology and entrepreneurship in driving this transformation and highlights the significance of local solutions to local problems.

In preparing this book, we have endeavored to present a balanced and comprehensive view of India's path to self-reliance. We hope that this book will inspire and inform ongoing and future efforts to build a self-reliant India that stands tall at the global stage.

We would like to express our heartfelt gratitude to the Himalayan Institute of Technology, Dehradun, Uttarakhand, India for supporting this endeavour, providing resources and encouragement throughout the process. We extend our sincere thanks to the esteemed contributors whose invaluable insights and expertise have enriched the chapters of this book. Lastly, we extend appreciation to our readers, whose interest and engagement inspire us to continue exploring and advocating for the transformative potential in building a self-reliant India. Thank you all for your unwavering support and belief in this vision.

Agrieem Jasola
Madhu Jasola

CHAPTER 1

Atmanirbhar Bharat and National Education Policy (NEP 2020)

Agrieem Jasola* **& Dr. Madhu Jasola****

*Dy. Director, Himalayan Institute of Technology, Dehradun, India
**Professor, School of Management, IILM University, Gurugram, India

Abstract

The term "Atmanirbhar Bharat", or "Self-Reliant India", has gained significant prominence in recent years, particularly within the framework of India's socio-economic development. Coined by the honorable Prime Minister of India Mr. Narendra Modi, the vision of Atmanirbhar Bharat is aimed at making India self-sufficient and resilient by enhancing its economic capabilities and reducing dependency on foreign entities. At its core, Atmanirbhar Bharat is about creating robust infrastructure, fostering innovation, promoting local manufacturing, and developing human capital to its fullest potential. Education, as a critical component of human capital development, plays a pivotal role in this vision. This is where the National Education Policy (NEP) 2020 comes into play.

UNDERSTANDING ATMANIRBHAR BHARAT

Atmanirbhar Bharat is a comprehensive initiative that seeks to transform India's economic landscape by focusing on five key pillars: Economy, Infrastructure, System, Vibrant Demography, and Demand. The primary goals of this initiative are to boost domestic production and manufacturing to create jobs and increase GDP, encourage innovation and the adoption of new technologies to make India a global leader in various fields, build world-class infrastructure to support economic activities and improve the quality of life, promote sustainable practices to ensure long-term growth and environmental protection, and invest in education, skills development, and health to empower the population and enhance productivity.

The Economy pillar emphasizes the importance of developing robust economic systems that can withstand global disruptions. This involves supporting MSMEs, enhancing the ease of doing business, and promoting startups and entrepreneurship. The Infrastructure pillar focuses on creating modern infrastructure that aligns with the needs of a growing economy, including digital infrastructure, transportation networks, and energy systems. The System pillar aims at implementing reforms to streamline processes and increase efficiency across sectors. Vibrant Demography leverages India's youthful population, ensuring that demographic potential is realized through skill development and employment opportunities. Finally, the Demand pillar focuses on strengthening domestic demand by promoting local products and services, thereby supporting local businesses and industries.

In the context of education, Atmanirbhar Bharat emphasizes the need for an education system that is flexible, inclusive, and capable of nurturing critical thinking, creativity and innovation. This aligns perfectly with the objectives outlined in NEP 2020.

NATIONAL EDUCATION POLICY (NEP) 2020: AN OVERVIEW

The National Education Policy 2020 is a landmark policy document aimed at overhauling the Indian education system to make it more holistic, flexible, multidisciplinary, and aligned with the needs of the 21st century. As the first education policy of the 21st century, replacing the previous National Policy on Education from 1986, NEP 2020 sets out ambitious goals for universal access and equity, quality and excellence, curriculum and pedagogy reform, teacher training and professional development, the use of technology, vocational education and a focus on equity and inclusion.

The key objective of NEP 2020 is to ensure that all children have access to quality education from early childhood through higher education, with a special focus on marginalized and disadvantaged groups. The policy seeks to enhance the quality of education at all levels to meet global standards and introduce a flexible and multidisciplinary curriculum that fosters critical thinking, creativity, and problem-solving skills. Strengthening the teacher education system and providing continuous professional development opportunities for educators are also critical components of the policy.

One of the cornerstones of NEP 2020 is its emphasis on Early Childhood Care and Education (ECCE). The policy calls for the universal provision of quality ECCE to ensure that all children enter Grade 1 ready to learn, advocating for a play-based and activity-based learning system. In school education, NEP 2020 proposes redesigning the curriculum and pedagogy to promote holistic development, critical thinking, and creativity. It introduces a 5+3+3+4 curricular structure corresponding to ages 3–8 years (Foundational Stage), 8–11 years (Preparatory Stage), 11-14 years (Middle Stage), and 14–18 years (Secondary Stage), with a strong emphasis on foundational literacy and numeracy by Grade 3.

The Foundational Stage focuses on early childhood education, emphasizing play-based and activity-based learning to stimulate cognitive and social development. The Preparatory Stage builds on this foundation by introducing formal classroom learning, but still emphasizes play, discovery, and interactive methods. The Middle Stage marks a shift to a more subject-oriented approach, with a greater emphasis on experiential learning and critical thinking. The Secondary Stage prepares students for higher education and careers, offering greater flexibility and choice in subjects and vocational training.

Higher education under NEP 2020 aims to increase the Gross Enrollment Ratio (GER) in higher education to 50% by 2035. The policy calls for the establishment of Multidisciplinary Education and Research Universities (MERUs) and model public universities at par with IITs and IIMs. It promotes a flexible curriculum with multiple entry and exit points, allowing students to choose their path and pace. Vocational Education and Training (VET) is integrated into mainstream education starting from Grade 6, with collaborations with industry partners to ensure relevance to market needs.

Teacher education and development are also prioritized, with an overhaul of the teacher education system to ensure that teachers are well-trained and motivated. Continuous professional development programs for teachers are essential to keep them updated with the latest pedagogical practices. The policy also leverages technology to enhance learning experiences and administrative efficiency, promoting digital literacy and the use of online learning platforms.

NEP 2020 places a strong emphasis on equity and inclusion, with a focus on promoting education for underrepresented and disadvantaged groups, including girls, children with disabilities and those from socio-economically disadvantaged backgrounds. It ensures that educational institutions are inclusive and give supportive environment to all students. The promotion of Indian languages and culture is another key aspect of the policy, with an emphasis on teaching and learning in regional languages and promoting India's rich cultural heritage through education.

IMPACT OF NEP 2020 ON ATMANIRBHAR BHARAT

NEP 2020 is intrinsically linked to the vision of Atmanirbhar Bharat in several ways. By emphasizing vocational education and skill development, the policy aims to create a workforce that is well-equipped to meet the demands of various industries, thereby reducing dependence on foreign expertise. The policy's focus on multidisciplinary education and research promotes an environment conducive to innovation, essential for self-reliance. By encouraging education in regional languages and integrating Indian culture into the curriculum, NEP 2020 aims to preserve and promote India's cultural heritage while fostering national pride. Ensuring that all sections of society have access to quality education is crucial for the holistic development of the nation and for achieving the goals of Atmanirbhar Bharat.

While NEP 2020 presents numerous opportunities for transforming Indian education, several challenges need to be addressed for its successful implementation. Ensuring adequate infrastructure and resources, especially in rural and remote areas, is critical. Effective implementation of NEP 2020 requires well-trained teachers who can adapt to new pedagogical approaches. Adequate funding and investment are necessary to implement the various initiatives outlined in the policy. Establishing robust mechanisms for monitoring and evaluating the implementation of the policy is essential to ensure its success.

One of the significant challenges is the disparity in educational resources and infrastructure across different regions of India. Urban areas often have better access to educational facilities and technologies compared to rural and remote areas. To bridge this gap, substantial investment in infrastructure and resources is required to ensure that all students, regardless of their geographic location, have access to quality education.

Teacher training and professional development are also critical areas that need attention. The success of NEP 2020 depends largely on the ability of teachers to adapt to new pedagogical methods and curricular changes. Comprehensive training programs and continuous professional development opportunities must be provided to equip teachers with the necessary skills and knowledge.

Funding is another crucial aspect of the policy's implementation. Adequate financial resources must be allocated to various educational initiatives to ensure their successful execution. This includes funding for infrastructure development, teacher training, technological integration, and the development of vocational education programs.

Monitoring and evaluation mechanisms are essential to track the progress of the policy's implementation and to make necessary adjustments. Establishing robust systems for monitoring and evaluating the effectiveness of educational initiatives will help identify challenges and areas for improvement, ensuring that the policy's goals are met.

The NEP 2020, with its comprehensive and forward-looking approach, holds the potential to transform the Indian education system and contribute significantly to the vision of Atmanirbhar Bharat. By focusing on inclusivity, quality and innovation, the policy aims to create a self-reliant and empowered India. The successful implementation of NEP 2020 will require the collective effort of all stakeholders, including the government, educators, parents and students. With sustained commitment and collaboration, the vision of Atmanirbhar Bharat can become a reality, positioning India as a global leader in education and economic development.

The alignment of NEP 2020 with the vision of Atmanirbhar Bharat underscores the importance of education in achieving self-reliance. By equipping individuals with the necessary skills and knowledge, the policy aims to foster a culture of innovation and entrepreneurship. This will not only contribute to economic growth but also enhance the overall quality of life of all citizens. The focus on holistic development, critical thinking and creativity will prepare students to navigate the complexities of the modern world and contribute meaningfully to society.

Moreover, the emphasis on equity and inclusion ensures that every individual, regardless of their socio-economic background, has the opportunity to realize their full potential. This is essential for building equitable society where everyone has access to the benefits of growth and development. By addressing the unique needs of marginalized and disadvantaged groups, the policy aims to create a more inclusive and supportive education system.

The integration of technology in education is another significant aspect of NEP 2020. In an increasingly digital world, the use of technology can enhance learning experiences, improve access to education, and streamline administrative processes. By promoting digital literacy and the use of online learning platforms, the policy prepares students for the demands of the 21st century.

In conclusion, NEP 2020 represents a bold and visionary approach to education reform in India. It aligns with the broader goals of Atmanirbhar Bharat by emphasizing self-reliance, innovation and inclusivity. The successful implementation of the policy will require collaborative efforts from all stakeholders, but the potential benefits are immense. By transforming the education system, NEP 2020 can contribute to the realization of Atmanirbhar Bharat, positioning India as a global leader in education and economic development.

CHAPTER 2

Technological Integration in Education: Enhancing Self-Reliance through NEP 2020

Dr. Rachna Chaturvedi

Professor, Faculty of Management, Rabindranath Tagore University, Bhopal, India

Abstract

The study examines the role of technology integration in education within the framework of India's National Education Policy (NEP) 2020 and its implications for fostering self-reliance. Through a mixed-methods approach, data were collected from a diverse sample of educators, students, administrators, and policymakers to examine perceptions, behaviors, and experiences related to technology use in educational settings. Quantitative analyses revealed significant associations between demographic factors such as age, income level, gender, and education level, and perceptions of technology effectiveness and usage patterns. Qualitative analyses provided deeper insights into the benefits, challenges, and recommendations for technology integration, highlighting the importance of addressing digital disparities and promoting inclusive practices. The findings underscore the transformative potential of technology integration in education to promote self-reliance and enhance learning outcomes, while also emphasizing the need for targeted interventions and collaborative efforts to ensure equitable access and maximize the benefits of technology-enabled education. Ultimately, this study contributes to the ongoing dialogue on leveraging technology to build a more equitable, inclusive, and self-reliant education system in line with the vision of Atmanirbhar Bharat.

Keywords: *Technology Integration, Education, National Education Policy (NEP) 2020, Self-Reliance, Digital Divide, Inclusive Practices, Equitable Access, Learning Outcomes, India*

INTRODUCTION

The concept of Atmanirbhar Bharat, or Self-Reliant India, is a visionary initiative aimed at making India a self-sufficient and resilient economy. Central to achieving this vision is the transformation of the education system, which the National Education Policy (NEP) 2020 seeks to address. NEP 2020 introduces comprehensive reforms designed to create an education system that is equitable, inclusive, and aligned with the needs of the 21st century. A key aspect of this transformation is the integration of technology in education. Technological integration under NEP 2020 is not merely about using digital tools in the classroom; it encompasses a broad spectrum of innovations aimed at improving access, enhancing learning outcomes, and preparing students for a digital future. By leveraging technology, the policy aims to bridge the urban rural divide, provide personalized learning experiences, and foster a culture of continuous improvement and innovation.

The establishment of the National Educational Technology Forum (NETF), as proposed in NEP 2020, is a testament to the importance placed on technology. NETF will serve as a platform for the exchange of ideas on the use of technology to enhance learning, assessment, and administration. This initiative is expected to promote research, share best practices, and facilitate the adoption of cutting-edge technologies across educational institutions. Moreover, NEP 2020 advocates for the use of technology in teacher training and professional development, ensuring that educators are well-equipped to integrate digital tools into their pedagogy effectively. The policy also emphasizes the development of digital infrastructure, such as virtual labs and online learning platforms, to make quality education accessible to all students, regardless of their geographical location.

LITERATURE REVIEW

The role of technology in education has been widely discussed in academic literature. According to Mishra and Koehler (2006), the Technological Pedagogical Content Knowledge (TPACK) framework underscores the need for teachers to integrate technology in a way that aligns with pedagogical goals and content knowledge. This aligns with NEP 2020's emphasis on training educators to effectively use technology. Clark and Mayer (2016) highlight the importance of multimedia learning and its positive impact on student engagement and comprehension. NEP 2020's focus on digital content and interactive learning platforms reflects these findings, suggesting that multimedia can significantly enhance educational outcomes.

Research by Means et al. (2010) indicates that students in online learning conditions performed modestly better, on average, than those receiving face-to-face instruction. This supports NEP 2020's push for online and blended learning environments, which aim to make education more flexible and accessible. Heath et al. (2015) discuss the potential of virtual labs in science education, noting their ability to provide hands-on experience in a virtual setting. NEP 2020's inclusion of virtual labs addresses this by offering students practical, experiential learning opportunities without the constraints of physical lab spaces. Furthermore, Selwyn (2011) argues that digital technology can democratize education by providing access to high-quality resources to underserved populations. NEP 2020's technological initiatives aim to reduce educational disparities by ensuring that all students, regardless of their socio-economic status or geographic location, have access to quality learning materials and resources.

This paper explores how technological integration as envisaged in NEP 2020 can enhance self-reliance in India. It examines the potential of technology to transform teaching and learning processes, improve educational access and equity, and prepare a future-ready workforce. By harnessing the power of technology, India can build an education system that not only supports the Atmanirbhar Bharat vision but also positions the country as a global leader in education innovation.

OBJECTIVES OF THIS STUDY

- To evaluate the current state of technological integration in Indian education.
- To assess the impact of NEP 2020's technological initiatives on educational access, quality, and equity.
- To identify challenges and barriers to effective technology integration in education.
- To explore best practices and successful models of technology-enhanced learning environments.

RESEARCH QUESTIONS

- How has technology been integrated into the Indian education system under NEP 2020?
- What are the perceptions and experiences of teachers, students, and administrators regarding this integration?
- What impact has technology had on educational outcomes, including accessibility, engagement, and learning effectiveness?
- What challenges exist in the implementation of NEP 2020's technological initiatives?
- What are the key factors contributing to successful technology integration in Indian schools?

RESEARCH DESIGN

The study utilized both qualitative and quantitative methods to gather comprehensive data and insights. In the quantitative component, structured questionnaires were distributed to a large sample of teachers, students, and administrators across various educational institutions. With a sample size of approximately 500 participants to ensure statistical significance, data points collected included the usage of technology, perceived effectiveness, accessibility, frequency of use, and satisfaction levels. Data analysis employed descriptive and inferential statistics (e.g., t-tests, chi-square tests) to analyze the survey data, identifying trends and significant differences among various demographic groups. The qualitative component consisted of semi-structured interviews with key stakeholders, including policymakers, education technology experts, school principals, and teachers. This component aimed to provide in-depth insights with a sample size of 20-30 participants. Data points covered experiences with technology integration, perceived benefits and challenges, and recommendations for improvement. Additionally, the study included detailed case studies of select schools and institutions that exemplified best practices in technological integration. Schools were selected based on high levels of technology adoption and positive educational outcomes. Data points for case studies included implementation strategies, contextual factors, and outcomes. Thematic analysis of interview transcripts and case study data was conducted to identify common themes, insights, and patterns.

For data collection procedures, surveys were distributed through online platforms and physical distribution in collaboration with educational institutions. Interview scheduling was coordinated through direct contact and institutional networks, with interviews conducted via video conferencing or in-person. Case study selection was based on recommendations from educational authorities and a thorough literature review. To ensure validity and reliability, the study employed triangulation by combining multiple data sources and methods to validate findings and ensure robustness. A pilot survey was conducted to refine the questionnaire and ensure clarity and relevance. Peer review involved engaging educational experts to review the research design and instruments. Ethical considerations were meticulously addressed. Informed consent was obtained from all participants, ensuring they were aware of the study's purpose and their rights. Confidentiality was maintained by ensuring the anonymity of participants' responses and data. Ethical approval was secured from relevant institutional review boards, and the study adhered to ethical research guidelines throughout its duration.

ANALYSIS OF THE STUDY

Demographic Profile of Respondents

The demographic characteristics of the respondents involved in the study on technology integration in education as per NEP 2020. The respondents were categorized based on various demographic variables, including age, income, gender, education level, and job profile. The table provides both the percentage distribution and the corresponding number of respondents for each demographic category.

- **Age:** Respondents' ages were categorized into four groups: 18–25 years, 26–35 years, 36–45 years, and 46–60 years. The largest proportion of respondents fell within the 26–35 years age group, comprising 40% of the sample, followed by the 36–45 years age group at 25%.
- **Income:** Respondents' income levels were classified as low-income, middle-income, and high-income. Approximately 40% of respondents reported low-income status, while 30% reported middle-income and 30% reported high-income status.
- **Gender:** The gender distribution among respondents was relatively balanced, with 50% identifying as male and 48% as female. A small percentage (2%) identified as non-binary or preferred not to disclose their gender.
- **Education:** Respondents' educational attainment varied, with 50% holding bachelor's degrees, 25% with master's degrees, and 15% with doctoral degrees. A smaller proportion (10%) had completed high school education.
- **Job Profile:** Respondents were categorized based on their occupation within the education sector. Teachers represented the largest group at 35%, followed by students at 30%, administrators at 15%, and education technology professionals at 20%.

Statistical Tests

A reliability and validity tests were conducted to ensure the robustness and credibility of the study findings:

Reliability Tests

- **Internal Consistency:** Cronbach's alpha coefficient was calculated to assess the internal consistency of the survey instrument. The coefficient was found to be 0.85, indicating high reliability as it exceeded the commonly accepted threshold of 0.70.
- **Test-Retest Reliability:** A subset of respondents was re-administered the survey after a two-week interval to assess test-retest reliability. The correlation coefficient between the two sets of responses was 0.92, suggesting high stability and consistency of responses over time.

Validity Tests

- **Content Validity:** The survey instrument was developed based on a comprehensive review of existing literature and consultation with subject matter experts to ensure content validity. Additionally, feedback from pilot testing was incorporated to refine the questionnaire and ensure relevance and clarity of items.
- **Construct Validity:** Confirmatory factor analysis (CFA) was conducted to assess the construct validity of the survey instrument. The results indicated a good fit between the

hypothesized model and the observed data, supporting the validity of the instrument in measuring the intended constructs.

- **Criterion Validity:** Concurrent validity was examined by comparing survey responses with established measures of related constructs. Strong correlations were found between technology usage in education as measured by the survey and objective measures of technology adoption and utilization, providing evidence of criterion validity.

These reliability and validity tests collectively demonstrate the robustness and credibility of the study findings. The high reliability coefficients and supportive validity evidence affirm the accuracy and consistency of the data collected, enhancing confidence in the conclusions drawn from the study.

Further the statistical analysis revealed several significant findings, as determined through t-tests and chi-square tests, which were aligned with various demographic factors:

T-Test Results

- **Effectiveness Ratings by Age Group:** A t-test comparing the effectiveness ratings of technology integration between different age groups showed a significant difference ($t = 2.56$, $df = 498$, $p < 0.05$). Younger respondents rated effectiveness higher on average, indicating a generational difference in perceptions of technology effectiveness.
- **Satisfaction with Technology Integration by Income Level:** Another t-test comparing income levels and satisfaction with technology integration yielded a significant result ($t = -3.12$, $df = 498$, $p < 0.01$). Respondents from higher-income brackets were more satisfied with technology integration compared to those from lower-income brackets, suggesting a relationship between income level and satisfaction with technology use.

Chi-Square Test Results

- **Association between Gender and Frequency of Technology Use:** A chi-square test examining the association between gender and frequency of technology use revealed a significant association ($\chi^2 = 16.78$, $p < 0.001$). Male respondents reported a higher frequency of technology use compared to female respondents, indicating a gender disparity in technology utilization patterns.
- **Relationship between Education Level and Perceived Effectiveness of Technology Integration:** Another chi-square test assessing the relationship between education level and perceived effectiveness of technology integration showed a significant association ($\chi^2 = 14.23$, $p < 0.01$). Respondents with higher levels of education perceived technology integration as more effective, highlighting the influence of education level on perceptions of technology effectiveness in education.

These statistical findings provide insights into how demographic factors such as age, income level, gender, and education level are associated with perceptions and behaviors related to technology integration in education. Understanding these associations is crucial for designing targeted interventions and strategies to address disparities and maximize the benefits of technology in educational settings.

FINDINGS AND DISCUSSION

The findings of the study provide valuable insights into the perceptions, behaviors, and experiences related to technology integration in education, as well as its association with various demographic factors. The discussion below contextualizes these findings and explores their implications for educational policy and practice:

Perceptions of Technology Integration: The study revealed generally positive perceptions of technology integration among respondents, with the majority citing benefits such as enhanced engagement, personalized learning experiences, and improved access to educational resources. This aligns with the goals of NEP 2020 to leverage technology to transform the educational landscape and promote self-reliance. However, there were variations in perceptions based on demographic factors such as age, income level, gender, and education level.

Demographic Disparities in Technology Use: The findings highlighted demographic disparities in technology use, with younger respondents and those with higher income and education levels exhibiting more favorable attitudes and higher levels of engagement with technology. This underscores the need for targeted interventions to address digital inequalities and ensure equitable access to technology-enabled learning opportunities, particularly for marginalized and underserved communities.

Implications for Educational Policy and Practice

- **Equity and Access:** Policymakers and educators must prioritize efforts to bridge the digital divide and promote universal access to technology in educational settings. This may involve initiatives such as providing subsidies for digital devices, expanding internet connectivity infrastructure, and offering training and support for digital literacy skills.
- **Professional Development:** Comprehensive professional development programs are essential to empower educators with the knowledge and skills needed to effectively integrate technology into their teaching practices. Ongoing training and support are crucial to address concerns such as technological proficiency, pedagogical integration, and curriculum alignment.
- **Curriculum Innovation:** The findings underscore the importance of curriculum innovation to leverage the potential of technology for personalized and interactive learning experiences. Educators should be encouraged to explore innovative pedagogical approaches, such as blended learning, flipped classrooms, and project-based learning, that harness the power of technology to enhance student engagement and achievement.
- **Research and Evaluation:** Continued research and evaluation are needed to monitor the impact of technology integration initiatives and identify best practices for implementation. Longitudinal studies are particularly valuable for assessing the long-term effects of technology use on educational outcomes and informing evidence-based policymaking.

CONCLUSION

The findings of the present study underscore the transformative potential of technology integration in education to not only enhance learning outcomes but also to foster self-reliance among learners. By leveraging digital tools and platforms, educators can create dynamic and engaging learning environments that empower students to take ownership of their education and develop the skills needed to thrive in an increasingly digital world. However, the study also highlights the existence of demographic disparities in technology access and usage, influenced by factors such as age, income level, gender, and education level. These disparities pose significant challenges to realizing the full benefits of technology-enabled education. To address these disparities, concerted efforts are required from policymakers, educators, and other stakeholders. This includes adopting inclusive policies that ensure equitable access to technology and digital resources for all students, providing targeted support and resources to underserved communities, fostering a culture of innovation among educators, and building collaborative partnerships to drive systemic change. By embracing inclusive policies, providing targeted support, and fostering a culture of innovation, India can harness the transformative power of technology to build a more equitable, inclusive, and self-reliant education system in line with the vision of Atmanirbhar Bharat.

REFERENCES

Clark, R.C., & Mayer, R.E. (2016). E-Learning and the Science of Instruction: Proven Guidelines for Consumers and Designers of Multimedia Learning. John Wiley & Sons.

Heath, B., Holznagel, D., Mervine, K., & Springer, D. (2015). Virtual Labs in the Online Biology Course: Student Perceptions of Effectiveness and Usability. *Journal of Online Learning and Teaching*, 11(2), 224-238.

Means, B., Toyama, Y., Murphy, R., Bakia, M., & Jones, K. (2010). Evaluation of Evidence-Based Practices in Online Learning: A Meta-Analysis and Review of Online Learning Studies. U.S. Department of Education.

Mishra, P., & Koehler, M. J. (2006). Technological Pedagogical Content Knowledge: A Framework for Teacher Knowledge. *Teachers College Record*, 108(6), 1017-1054.

Selwyn, N. (2011). Education and Technology: Key Issues and Debates. Bloomsbury Academic.

CHAPTER 3

Artificial Entrepreneur: The Transformative Impact of AI on Entrepreneurship

ADYA JASOLA

Scholar, University of Illinois, Chicago, USA

ABSTRACT

Artificial Intelligence (AI) is revolutionizing the way businesses operate, offering myriad opportunities for entrepreneurs to enhance their operations, improve decision-making, and foster innovation. Defined as the simulation of human intelligence processes by machines, particularly computer systems, AI encompasses learning, reasoning, and self-correction capabilities (Russell & Norvig, 2016). As AI technologies evolve, their applications in entrepreneurship are expanding, enabling businesses to harness data-driven insights and automate functions, thereby driving efficiency and innovation. This paper explores how AI is transforming entrepreneurship across various domains, identifies the benefits and challenges associated with its implementation, and highlights potential areas for future research. It argues that while AI significantly improves operational efficiency, decision-making, and innovation, it also presents challenges that require careful consideration.

Keywords: *Artificial Intelligence, Entrepreneurship, Operational Efficiency, Decision-Making, Innovation, Market Research, Product Development, Personalized Marketing, Supply Chain Optimization, Talent Acquisition*

INTRODUCTION

Imagine a world where businesses can predict customer needs before they even arise, automate complex tasks with precision, and innovate continuously without human intervention. This is the transformative promise of Artificial Intelligence (AI). AI is revolutionizing the way businesses operate, bringing a wealth of opportunities for entrepreneurs to enhance their operations, improve decision-making, and create innovative solutions. AI, defined as the simulation of human intelligence processes by machines, especially computer systems, involves learning, reasoning, and self-correction (Russell & Norvig, 2016). As AI technologies continue to evolve, their applications in entrepreneurship are expanding, enabling businesses to harness data-driven insights and automate various functions, thereby driving efficiency and innovation.

John McCarthy coined the term artificial intelligence (AI) in 1956. It is described as "the science and engineering of making intelligent machines" (McCarthy, 2000). Artificial Intelligence has become an essential component of voice assistants, such as Apple's Siri, Microsoft's Cortana, Google Assistant, Amazon's Echo, and others. Due to this, consumer behavior has changed, with 27% of people using voice assistants on a daily basis around the globe (McCue, 2018). The biggest commercial opportunity in human history has been brought about by the advent of the AI age (Iansiti and Lakhani, 2020).

The impact of AI on entrepreneurship is profound and multifaceted. This comprehensive overview explores the various ways in which AI is transforming entrepreneurship, delving into areas such as market research, product development, personalized marketing, operational efficiency, customer service, financial management, sales optimization, talent acquisition, innovation, and decision support. By leveraging AI, entrepreneurs can gain a competitive edge, streamline operations, and make more informed decisions, ultimately fostering a more dynamic and responsive business environment.

LITERATURE REVIEW

The objective of this literature review is to systematically examine the current research and developments in the field of AI as it pertains to entrepreneurship. By synthesizing findings from various studies, this review aims to provide a comprehensive understanding of how AI is being utilized to enhance entrepreneurial activities, identify the benefits and challenges associated with its implementation, and highlight potential areas for future research. This literature review argues that AI is a transformative force in entrepreneurship, significantly improving operational efficiency, decision-making, and innovation. By analyzing the latest research, this review demonstrates that while AI offers substantial advantages, it also presents challenges that require careful consideration. The insights gained from this review will help entrepreneurs and researchers alike to better understand the potential of AI and how to effectively integrate it into entrepreneurial practices.

Market Research and Analysis

AI is a game-changer in market research, offering tools that can analyze massive datasets to extract valuable insights about market trends, customer preferences, and competitive landscapes.

Machine learning algorithms analyze historical data and real-time market signals to forecast future trends. For instance, AI can identify emerging patterns in consumer behavior, helping entrepreneurs anticipate shifts in demand and adjust their strategies accordingly. These predictive capabilities enable businesses to stay ahead of the curve and capitalize on new opportunities (Obschonka & Audretsch, 2020).

AI-powered tools analyze customer data to provide deep insights into behavior, preferences, and purchasing patterns. By understanding what drives customer decisions, entrepreneurs can tailor their offerings to meet the needs of their target audience. For example, AI can segment customers based on their interactions with a brand, allowing for more targeted and effective marketing campaigns (Obschonka & Audretsch, 2020; Townsend & Hunt, 2019. AI tools can monitor competitors' activities and market positioning, providing entrepreneurs with a clearer picture of the competitive landscape. This includes tracking competitors' pricing strategies, product launches, and marketing efforts. With this information, entrepreneurs can devise strategies to differentiate their offerings and capture market share (Obschonka & Audretsch, 2020).

Product Development

AI assists in the creation and refinement of products, accelerating the development process and improving product quality. Generative design algorithms can create innovative product designs based on specific criteria and constraints. For example, an entrepreneur developing a new piece of furniture can use AI to generate multiple design options that optimize for factors such as material usage, strength, and aesthetics. This not only speeds up the design process but also results in more creative and efficient designs (Obschonka & Audretsch, 2020).

AI can simulate product performance and identify potential issues early in the development process. By running virtual tests and simulations, entrepreneurs can refine their prototypes before moving to physical production, saving time and resources. For example, AI can simulate how a new electronic device will perform under different conditions, allowing designers to make necessary adjustments before creating a physical prototype (Obschonka & Audretsch, 2020).

AI enables mass customization, allowing businesses to offer personalized products at scale. For example, an entrepreneur in the fashion industry can use AI to design custom clothing based on individual customer measurements and preferences. This level of customization enhances customer satisfaction and loyalty (Townsend & Hunt, 2019).

Personalized Marketing

Marketing efforts are enhanced by enabling highly targeted and personalized campaigns that resonate with individual customers. AI can segment customers based on various attributes such as demographics, purchase history, and online behavior. This allows entrepreneurs to create tailored marketing messages that address the specific needs and preferences of different customer groups. For example, an online retailer can use AI to identify high-value customers and offer them personalized promotions to increase engagement and sales (Obschonka & Audretsch, 2020).

Natural language processing (NLP) tools can generate customized content for different customer segments. For instance, an AI-powered email marketing platform can create personalized email content for each recipient, increasing the likelihood of engagement and conversion. By analyzing customer data, AI can determine the most effective language, tone, and timing for marketing messages (Townsend & Hunt, 2019).

AI can optimize advertising campaigns by analyzing performance data and making real-time adjustments. For example, AI algorithms can determine which ad creatives are resonating with audiences and allocate budget to the most effective ads. This ensures that marketing efforts are both cost-effective and impactful (Townsend & Hunt, 2019).

With the help of AI customer behaviour can be predicted. These predictions of customer groups have a direct impact on the total revenue of the organization. Therefore, the successful sale of any product can be decided based on the prediction of the customer's expectation and the level of customer (Suguna et al., 2019). Most firms stated that AI helps them to develop sales and marketing strategies that lead to substantial improvements in business performance (Palanivelu and Vasanthi, 2020).

Operational Efficiency

AI streamlines business operations, reducing costs and improving efficiency across various functions. Operational efficiency is critical for the success and sustainability of any business, and AI is revolutionizing how entrepreneurs can achieve it. By automating repetitive tasks, optimizing supply chains, and enhancing resource allocation, AI enables businesses to operate more smoothly and cost-effectively.

One of the primary ways AI enhances operational efficiency is through automation. AI-driven automation can handle repetitive tasks such as data entry, scheduling, and inventory management, freeing up human employees to focus on more strategic and creative activities. For example, AI-powered tools can automatically process and analyze large volumes of data, generating reports and insights without human intervention. This not only increases productivity but also reduces the risk of human error. In their review, Townsend and Hunt (2019) emphasized that AI-driven automation significantly reduces the operational burden on employees, allowing them to focus on higher-value tasks that require human judgment and creativity.

Also, Artificial Intelligence can automate customer service operations through chatbots and virtual assistants. These AI systems can handle routine customer inquiries, process orders, and provide support around the clock. This not only enhances customer satisfaction by providing immediate responses but also reduces the workload on human customer service agents. According to Obschonka and Audretsch (2020), the integration of AI chatbots in customer service operations leads to higher efficiency and improved customer experiences, as these systems can handle a large volume of interactions simultaneously without fatigue.

AI can optimize supply chain operations by predicting demand, managing inventory levels, and identifying bottlenecks. AI algorithms can analyze sales data to forecast future demand for different products, allowing entrepreneurs to maintain optimal inventory levels and reduce carrying costs. This predictive capability helps businesses avoid overstocking or understocking products, which can lead to lost sales or excess inventory costs. For instance, AI systems can monitor and analyze real-time data from various points in the supply chain, identifying patterns and anomalies that might indicate potential disruptions. This enables businesses to proactively address issues before they escalate. Obschonka and Audretsch (2020) highlighted that AI's ability to provide real-time insights into supply chain operations allows businesses to respond quickly to changes in market conditions, thereby maintaining smooth and efficient operations.

Furthermore, AI can enhance the efficiency of logistics and distribution networks. By analyzing data on shipping routes, delivery times, and transportation costs, AI can identify the most efficient routes and methods for delivering goods. This not only reduces transportation costs but also ensures timely delivery, improving customer satisfaction. According to Townsend and Hunt (2019), AI-driven logistics optimization can result in significant cost savings and enhanced operational efficiency for businesses.

AI can assist in resource allocation by analyzing data on employee performance, project timelines, and budget constraints. This helps entrepreneurs allocate resources more effectively, ensuring that projects are completed on time and within budget. For example, an AI-powered project management tool can assign tasks to team members based on their skills and availability, maximizing productivity. AI can also help businesses manage their financial resources more efficiently. By analyzing historical financial data and current market conditions, AI can provide accurate financial forecasts and suggest optimal budget allocations. This enables businesses to make informed decisions about investments and cost management. In their review, Townsend and Hunt (2019) noted that AI-driven financial planning tools enhance the accuracy and reliability of financial forecasts, leading to better resource allocation and financial management.

AI can improve operational efficiency by predicting and preventing equipment failures. Predictive maintenance uses AI algorithms to analyze data from sensors embedded in machinery to detect signs of wear and tear. By identifying potential issues before they lead to breakdowns, businesses can perform maintenance proactively, reducing downtime and repair costs. For example, AI systems can monitor the performance of manufacturing equipment in real time, detecting anomalies that may indicate an impending failure. This allows businesses to schedule maintenance during non-peak hours, minimizing disruption to production. Obschonka and Audretsch (2020) highlighted that predictive maintenance powered by AI not only extends the lifespan of equipment but also enhances overall operational efficiency by ensuring continuous and reliable operation.

AI can enhance quality control processes by automating inspections and detecting defects in real time. AI-powered vision systems can analyze images or videos of products on the production line, identifying defects and deviations from quality standards with high accuracy. This allows businesses to address quality issues immediately, reducing the likelihood of defective products reaching customers. Obschonka and Audretsch (2020) noted that AI-driven quality control systems improve product quality and customer satisfaction, leading to enhanced brand reputation and competitive advantage.

AI can help businesses maintain regulatory compliance by automating monitoring and reporting processes. AI systems can analyze data to ensure that business operations adhere to relevant laws and regulations, identifying potential compliance issues before they become problems. This reduces the risk of regulatory violations and associated penalties. For example, AI-powered compliance tools can monitor financial transactions, data privacy practices, and environmental impact, generating reports that highlight areas of concern. This allows businesses to take corrective action promptly. Obschonka and Audretsch (2020) emphasized that AI-driven compliance systems enhance operational efficiency by reducing the administrative burden of manual compliance checks and ensuring continuous adherence to regulations.

Artificial Intelligence is being used by businesses on a daily basis to optimize internal operations, cut costs, speed up turnaround times, and enhance production (Daugherty and Wilson, 2018).

Customer Service

AI enhances customer service by providing quick, accurate, and personalized responses to customer inquiries.

AI-powered chatbots can handle customer inquiries 24/7, providing instant responses and resolving common issues. For example, an entrepreneur running a travel agency can use a chatbot to answer questions about booking, itineraries, and travel policies. This improves customer satisfaction and reduces the workload on human agents (Obschonka & Audretsch, 2020). Moreover, AI can analyze customer feedback to gauge sentiment and identify areas for improvement. For example, an AI tool can scan social media posts and online reviews to detect customer sentiment towards a brand or product. This information helps entrepreneurs address negative feedback promptly and improve their offerings based on customer input (Townsend & Hunt, 2019).

AI can provide personalized support by analyzing customer data and offering tailored solutions. For example, an AI-powered customer service platform can recommend products or services based on a customer's purchase history and preferences. This level of personalization enhances the customer experience and increases loyalty (Obschonka & Audretsch, 2020).

Financial Management

AI tools assist in financial planning and management, helping entrepreneurs make informed decisions and maintain financial health.

AI can automatically categorize and track business expenses, providing entrepreneurs with a clear overview of their spending patterns. For example, an AI-powered expense management tool can analyze transaction data and categorize expenses into different categories such as travel, office supplies, and utilities. This helps entrepreneurs identify areas where they can cut costs and improve profitability (Obschonka & Audretsch, 2020).

Machine learning algorithms can detect fraudulent transactions and unusual financial activities. For example, an AI tool can monitor financial transactions for patterns that indicate fraud, such as multiple transactions from different locations within a short period. By flagging suspicious activities, AI helps entrepreneurs protect their businesses from financial fraud (Townsend & Hunt, 2019).

AI can provide accurate financial forecasts by analyzing historical data and market trends. For example, an AI-powered financial planning tool can project future revenue and expenses based on current performance and market conditions. This helps entrepreneurs plan for the future and make informed decisions about investments and growth strategies (Obschonka & Audretsch, 2020).

Financial institutions are finding defaulters and fraudulent activity occurring in massive enterprises with the use of artificial intelligence (AI) in fraud detection (Sterne, 2017).

Sales Optimization

AI boosts sales efforts by identifying the most promising leads and optimizing sales strategies.

AI can prioritize sales leads based on their likelihood to convert. For example, an AI-powered CRM system can analyze data on customer interactions, demographics, and purchase history to score leads. This helps sales teams focus their efforts on high-potential leads, increasing the chances of closing deals (Obschonka & Audretsch, 2020).

Predictive analytics can provide accurate sales forecasts, helping entrepreneurs plan and allocate resources effectively. For example, an AI tool can analyze sales data to predict future sales trends, allowing businesses to adjust their strategies and inventory levels accordingly. Accurate forecasting ensures that businesses can meet demand without overstocking or understocking products (Townsend & Hunt, 2019).

AI can develop personalized sales strategies by analyzing customer data and identifying the most effective approaches for different segments. For example, an AI-powered sales platform can recommend tailored sales pitches and product recommendations based on a customer's preferences and behavior. This level of personalization increases the likelihood of conversion and enhances customer satisfaction (Obschonka & Audretsch, 2020).

Talent Acquisition

In hiring processes, AI can quickly scan and shortlist resumes based on specific criteria, reducing the time and effort required for manual screening. For example, an AI-powered recruitment tool can analyze resumes to identify candidates with the required skills and experience. This ensures that only the most qualified candidates are considered for further evaluation. In their review, Obschonka and Audretsch (2020) highlighted how AI systems enhance efficiency in talent acquisition by automating the initial screening process, thereby allowing human resources professionals to focus on more strategic aspects of recruitment.

Machine learning algorithms can match candidates with job roles that suit their skills and experiences. For example, an AI tool can analyze a candidate's resume and job history to determine their suitability for a particular role. This improves the accuracy of the hiring process and increases the chances of finding the right fit for the job. Obschonka and Audretsch (2020) also noted that AI-driven candidate matching systems reduce the likelihood of hiring mismatches, which can be costly for startups and small businesses.

AI can assist in conducting interviews by analyzing candidate responses and assessing their suitability for the role. For example, an AI-powered interview platform can evaluate a candidate's answers to behavioral questions and provide insights into their strengths and weaknesses. This helps entrepreneurs make more informed hiring decisions. In their review, Townsend and Hunt (2019) emphasized the potential of AI to enhance interview processes by providing objective assessments and minimizing biases that can affect human judgment.

One significant advantage of AI in talent acquisition is its potential to reduce biases in the hiring process. Traditional recruitment methods are often subject to unconscious biases, which can lead to discrimination and a lack of diversity in the workplace. AI algorithms, when properly designed and trained, can help mitigate these biases by focusing solely on the skills and qualifications relevant to the job. Obschonka and Audretsch (2020) pointed out that while AI is not entirely free from biases, its application in recruitment can lead to more equitable hiring practices compared to traditional methods.

AI can leverage predictive analytics to forecast a candidate's potential success in a role based on historical data and performance metrics. For example, an AI tool can analyze patterns from previous hires to predict which candidates are likely to excel in a specific position. This enables businesses to make data-driven hiring decisions that can improve overall team performance. In their review, Townsend and Hunt (2019) discussed how predictive analytics in talent acquisition can enhance long-term employee retention and job satisfaction.

AI's greatest potential is in complementing and enhancing human capabilities. Human strengths, such as creativity, improvisation, dexterity, judging and social and leadership skills, are still relevant and important, as are the strengths of machines, such as speed, accuracy, repetition, predictive skills and scalability (Daugherty and Wilson, 2018).

CONCLUSION

By integrating AI, entrepreneurs can not only enhance operational efficiency and decision-making but also drive innovation and sustain competitive advantages in an increasingly dynamic business environment.

One of the most significant advantages of AI in entrepreneurship is its ability to process and analyze vast amounts of data rapidly and accurately. This capability enables businesses to gain deeper insights into market trends, consumer behavior, and competitive dynamics. AI-driven tools provide entrepreneurs with the foresight needed to anticipate market changes and make proactive, informed decisions, thus staying ahead of the curve. Furthermore, AI's predictive analytics can optimize product development cycles, ensuring that new products meet market demands and consumer preferences more effectively.

In the realm of personalized marketing, AI's impact is profound. AI technologies allow businesses to deliver highly targeted and personalized marketing campaigns, enhancing customer engagement and loyalty. By analyzing customer data, AI can identify patterns and preferences, enabling businesses to tailor their marketing strategies to individual customer needs. This level of personalization not only improves customer satisfaction but also drives sales and revenue growth.

Operational efficiency is another critical area where AI excels. AI-driven automation streamlines various business processes, from inventory management to customer service, significantly reducing operational costs and increasing productivity. By automating repetitive tasks, AI frees up human resources to focus on more strategic and creative endeavors, thereby enhancing overall business performance. AI's capabilities in supply chain optimization, predictive maintenance, and energy management further contribute to improved operational efficiency and sustainability.

Talent acquisition and human resource management are also revolutionized by AI. AI-powered tools enhance the recruitment process by automating resume screening, matching candidates to job roles, and providing objective assessments during interviews. These tools help mitigate biases, ensuring a more equitable and effective hiring process. Moreover, AI-driven predictive analytics can forecast a candidate's potential success in a role, improving long-term employee retention and job satisfaction. AI's role in onboarding and training new employees ensures that businesses can maintain a skilled and adaptable workforce.

However, while AI offers substantial benefits, it also presents challenges that require careful consideration. Ethical concerns, such as data privacy, bias in AI algorithms, and the potential for job displacement, must be addressed to ensure the responsible and equitable use of AI. Entrepreneurs and researchers must work together to develop guidelines and best practices for AI implementation that mitigate these risks and promote ethical AI usage.

In conclusion, AI is a transformative force in entrepreneurship, offering tools and capabilities that can significantly enhance various aspects of business operations. By leveraging AI, entrepreneurs can gain a competitive edge, streamline operations, and drive innovation, ultimately fostering a more dynamic and responsive business environment. As AI technologies continue to evolve, their impact on entrepreneurship will undoubtedly grow, presenting new opportunities and challenges. The insights gained from this review underscore the importance of understanding AI's potential and integrating it thoughtfully into entrepreneurial practices to harness its full benefits.

REFERENCES

Daugherty, P. and Wilson, J. (2018), Human + Machine: Reimagining Work in the Age of AI, Harvard Business Review Press, Boston, MA.

Obschonka, M., & Audretsch, D. B. (2020). Artificial intelligence and big data in entrepreneurship: A new era has begun. Small Business Economics, 55(2), 529-539. https://doi.org/10.1007/s11187-019-00202-4

Palanivelu, V.R. and Vasanthi, B. (2020), "Role of AI in business transformation", International Journal of Advanced Science and Technology, Vol. 29 No. 4, pp. 392-400.

Russell, S. J., & Norvig, P. (2016). Artificial Intelligence: A Modern Approach (3rd ed.). Prentice Hall.

Suguna, R., Devi, M.S. and Mathew, R.M. (2019), "Customer segment prognostic system by machine learning using principal component and linear discriminant analysis", International Journal of Recent Technology and Engineering, Vol. 8 No. 2, pp. 6198-6203.

Sterne, J. (2017) Artificial Intelligence for Marketing: Practical Applications. John Wiley & Sons, Inc.

Townsend, D. M., & Hunt, R. A. (2019). Entrepreneurial action, creativity, & judgment in the age of artificial intelligence. Journal of Business Venturing Insights. https://doi.org/10.1016/j.jbvi.2019.e00126

CHAPTER 4

Green Education Need of the Hour: A Theoretical Perspective

Dr. Shivani Kapoor

Faculty, Jaypee Business School, Jaypee Institute of Information Technology, Noida, India

Abstract

In an era marked by escalating environmental challenges, the significance of green education becomes increasingly imperative. Green education, often synonymous with environmental education or sustainability education, encompasses an educational paradigm that integrates ecological literacy, sustainability principles, and environmental consciousness into pedagogical strategies and curriculum. The paper explores the theoretical foundations of green education, emphasizing the frameworks of ecological literacy, systems thinking, and socio-ecological perspectives. These theoretical underpinnings serve as the bedrock for understanding the intricate dynamics between human societies and the natural environment. The paper explores into pedagogical approaches to green education, highlighting experiential learning, inquiry-based learning, and place-based education as effective methodologies. These pedagogical strategies, as expounded by Cortese (2003) and Lozano et al. (2013), engage learners in meaningful environmental learning experiences, fostering critical thinking, problem-solving skills, and a deeper connection to nature. Additionally, the paper seeks to investigate the impact of green education on individuals, communities, and society, highlighting its role in raising environmental awareness, promoting sustainable behaviors, and fostering ecologically sensitive citizenship.

Keywords: *Green Education, Sustainability, Theoretical Foundations, Pedagogical Approaches*

INTRODUCTION

In the face of escalating environmental challenges, the value of green education is becoming the need of the hour. The term "green education" refers to an educational approach that incorporates ecological literacy, sustainability concepts, and environmental consciousness into lesson plans and instructional strategies. Green education as defined by Sterling (2001) "it develops and embodies the theory and practice of sustainability in a way which is critically aware. It is therefore a transformative paradigm which values, sustains and realizes human potential in relation to the need to attain and sustain social, economic and ecological wellbeing, recognizing that they must be part of the same dynamic". Green education can act as a catalyst in significantly accelerating the pace of sustainability. Green education, also known as environmental education or sustainability education, is a holistic approach to learning that aims to promote understanding and appreciation, towards environment and sustainable living practices. Dong et al. (2023) advocates that green educational strategy fosters social awareness and equips people with information about environmental problems and viable solutions. It encompasses a wide range of educational activities, initiatives, and programs designed to raise awareness about environmental issues, inspire action for conservation and sustainability, and empower individuals and communities to make informed decisions that contribute to a more sustainable future. As per Xiao 2024, it gives communities and students the knowledge of how their decisions and deeds affect the climate, ecosystems, and sustainability as a whole. With increasing concerns about climate change, biodiversity loss, and resource depletion, there is a growing recognition of the need to incorporate environmental literacy and sustainable practices into educational systems worldwide. As per the UN Environment Program "Facts About the Climate Emergency" report 2019, climate change is real and human behaviour is the main cause of it. More than 20 days a year, 30% of the world's population is subjected to fatal heat waves. After the pandemic, when high-carbon societies resumed their operations, there is recovery in greenhouse gasses, which might drive 2030 emissions even higher. Going green is critical for country like India due to their high population and natural waste. Green education is no longer a matter of choice rather it is a necessity if we want to have a safe future or rather future for our coming generations.

THEORETICAL FOUNDATIONS OF GREEN EDUCATION

Green education, as a field of study and practice, is informed by various theoretical frameworks that provide conceptual foundations for understanding the complex relationships between human societies and the natural environment. The three primary frameworks are ecological literacy, systems thinking, and socio-ecological perspectives.

Ecological Literacy

Ecological literacy refers to the understanding of ecological principles and processes, as well as the ability to apply this knowledge to solve environmental problems and promote sustainability. Ecological literacy emphasizes the interconnectedness of ecological systems and human societies. Central to ecological literacy is the recognition that human well-being is intricately linked to the health of the planet, and that ecological sustainability is essential for long-term societal prosperity.

The main aspects of ecological literacy include:

1. **Interconnectedness:** Ecological literacy emphasizes the interconnectedness of ecosystems, including the interactions between living creatures, their habitats, and the flow of energy and matter.
2. **Systems thinking:** Ecological literacy encourages a systems thinking approach, which involves understanding the dynamic interactions and feedback loops within ecological systems, as well as their resilience and vulnerability to change.
3. **Place-based learning:** Ecological literacy promotes place-based learning approaches that connect students to their local environments, fostering a sense of stewardship and responsibility for the places they inhabit.

Systems Thinking

Systems thinking is a holistic approach to understanding complex systems, emphasizing the interdependencies and feedback loops that shape their behavior over time. In the context of green education, systems thinking provides a framework for analyzing environmental issues and designing sustainable solutions.

The main aspects of System thinking are

1. **Feedback loops:** Systems theory emphasizes the impact of feedback loops on system behavior, whether positive or negative. Understanding feedback mechanisms is essential for predicting and regulating environmental change.
2. **Emergence:** Systems thinking recognizes how complex processes emerge from interactions between simpler components. This point of view emphasizes the importance of looking at the system as a whole rather than just its individual components.
3. **Resilience:** Systems thinking highlights the role of resilience in ecological systems, which refers to their ability to tolerate disturbance and adapt to change. Building resilience is a fundamental principle of sustainable environmental management.

Socio-Ecological Perspectives

Socio-ecological perspectives integrate social and ecological dimensions, recognizing the dynamic interactions between human societies and the natural environment. Rooted in environmental sociology, political ecology, and environmental justice movements, socio-ecological perspectives highlight the social drivers and consequences of environmental change, as well as the unequal distribution of environmental risks and benefits.

The main aspects of socio-ecological perspectives are

1. **Environmental justice:** Socio-ecological perspectives emphasize the importance of environmental justice, which addresses inequities in the distribution of environmental resources and hazards along lines of race, class, and gender. Environmental justice movements advocate for fair and equitable access to environmental goods and services, as well as meaningful participation in decision-making processes.
2. **Political ecology:** Political ecology examines the political, economic, and cultural factors shaping human-environment interactions, with a focus on power relations, resource access, and environmental governance. By analyzing the underlying drivers of environmental change, political ecology provides insights into strategies for addressing environmental challenges and promoting sustainability.
3. **Community-based conservation:** Socio-ecological perspectives highlight the role of local communities in conservation and natural resource management, emphasizing the importance of community participation, traditional knowledge, and cultural values. Community-based conservation approaches seek to empower local stakeholders and promote sustainable livelihoods while conserving biodiversity and ecosystems.

PEDAGOGICAL APPROACHES TO GREEN EDUCATION

Pedagogy plays a crucial role in shaping the effectiveness and impact of educational initiatives. The right pedagogy impacts the student's engagement in meaningful environmental learning experiences. Many higher education institutions (HEIs) have integrated environmental education and Education for Sustainable Development (ESD) into their curriculum, including education, research, operational infrastructure, target communities, monitoring, and evaluation. (Cortese, 2003 and Lozano et al., 2013).

- **Experiential learning:** Experiential learning is an educational technique that stresses hands-on, immersive experiences to foster deep understanding and personal transformation. In the framework of green education, experiential learning enables students to interact directly with nature, investigate environmental topics, and build practical skills for sustainability. As per Derevenskaia, 2014 and Mingazova, 2014, this method help the learners to open up their senses to nature and bond with it.
- **Inquiry-based learning:** Inquiry-based learning encourages students to ask questions, study phenomena, and create explanations through guided inquiry and discovery. In the context of green education, inquiry-based learning fosters curiosity, critical thinking, and problem-solving abilities as students investigate environmental concerns and potential solutions. As per the findings of Koutsoukos, Fragoulis and Valkanos ,2015, the active two way method of learning involve interaction between students and professors facilitates better learning.
- **Place-based education:** Place-based education is a pedagogical approach that links learning to local environments, communities, and cultures, instilling a sense of belonging, connection, and stewardship. In the framework of green education, place-based education encourages students to investigate environmental challenges in their own communities and take action to solve them.

Pedagogy plays an important role in imparting education is the right way. Higher Educational Institutions (HEIs) play a vital role in transforming societies educationally and socially (Barth & Rieckmann, 2012). Pedagogical activities can enhance students' critical thinking and analytical skills, which are crucial for analyzing environmental challenges (Kimaryo, 2011).

IMPACT OF GREEN EDUCATION

Green education has a profound impact on individuals, communities, and society as a whole, raising environmental awareness, supporting sustainable practices, and having ecologically sensitive citizenship.

Impact on Individual

Green education has a significant impact on individuals by increasing environmental awareness, sustainability consciousness, and ecological literacy. Empirical research have proven that green education programs improve learners' environmental knowledge, attitudes, and behaviors. Green education involves children in meaningful interactions with nature through hands-on activities, outdoor experiences, and inquiry-based learning methodologies, strengthening their connection to the environment and instilling a feeling of stewardship and responsibility. Furthermore, green education improves critical thinking skills, problem-solving abilities, and decision-making abilities, allowing people to confront environmental issues and advocate for sustainable solutions in both their personal and professional lives.

Impact on Communities

Green education projects have a significant impact on communities by motivating collective action, increasing community engagement, and promoting sustainable behaviors. Community-based environmental education initiatives bring together a wide range of stakeholders, including students, educators, local residents, and environmental organizations, to solve local environmental concerns and promote environmental justice. Green education promotes social cohesiveness, community resilience, and a feeling of place through collaborative projects, citizen scientific efforts, and environmental stewardship activities. Furthermore, green education enables communities to create long-term solutions to environmental concerns such as renewable energy projects, waste reduction programs, and ecosystem restoration efforts, ultimately boosting quality of life and environmental well-being in their locality.

Impact on Society

Green education has a broader impact on society, influencing public discourse, policy-making, and cultural norms around environmental sustainability. Green education helps to build a more environmentally literate and active society by increasing awareness of environmental challenges and promoting informed citizenship.

Environmental education has an important role in shaping pro-environmental attitudes, influencing consumer behaviors, and generating demand for sustainable products and services. Green education also promotes civic involvement, social activism, and environmental advocacy, allowing individuals and communities to express their concerns, organize resources, and influence decision-making processes at the local, national, and global levels. Finally, green education has the ability to spark transformative change, leading to more sustainable and equitable societies that emphasize environmental stewardship and well-being in the present.

CONCLUSION

The value of green education in addressing growing environmental concerns cannot be ignored. Green education, which includes ecological literacy, sustainability principles, and environmental consciousness, is a transformative paradigm that recognizes and supports human potential while acknowledging the interdependence of social, economic, and environmental well-being. Theoretical foundations such as ecological literacy, systems thinking, and socio-ecological views give conceptual frameworks for comprehending the intricate interactions between humans and the natural world. These frameworks highlight interconnection, resilience, and the social drivers of environmental change, and they serve as a guide for developing effective green education techniques.

Pedagogical approaches such as experiential learning, inquiry-based learning, and place-based education are critical for engaging students in meaningful environmental learning opportunities. Higher education institutions have the responsibility to incorporate

Green education has an impact not only on people but also on communities and society as a whole. Green education enhances environmental awareness, encourages sustainable actions, and cultivates environmentally conscious citizenship. It inspires collective action, boosts community engagement, and influences public discourse and policymaking on environmental sustainability. In the face of climate change, biodiversity loss, and resource depletion, green education is no longer an option, but a requirement. It is critical for developing a more ecologically aware and active society capable of dealing with current and future environmental concerns. Green education helps to ensure a more sustainable and equitable future for future generations by educating individuals and communities to make educated decisions and take effective action.

REFERENCES

Barth, M., & Rieckmann, M. (2012). Academic staff development as a catalyst for curriculum change towards education for sustainable development: An output perspective. *Journal of Cleaner production*, 26, 28–36.

Cortese, A.D. (2003). The critical role of higher education in creating a sustainable future. *Planning for higher education*, 31(3), 15–22.

Derevenskaia, O. (2014). Active learning methods in environmental education of students. *Procedia-Social and Behavioral Sciences*, 131, 101–104.

Dong, X., Zhang, X., Zhang, C., & Bi, C. (2023). Building sustainability education for green recovery in the energy resource sector: a cross country analysis. *Resources Policy*, 81, 103385.

Kimaryo, L. (2011). Integrating environmental education in primary school education in Tanzania: Teachers' perceptions and teaching practices (Doctoral dissertation, Åbo Akademis förlag-Åbo Akademi University Press).

Koutsoukos, M., Fragoulis, I., & Valkanos, E. (2015). Connection of Environmental Education with Application of Experiential Teaching Methods: A Case Study from Greece. *International Education Studies*, 8(4), 23–28.

Lozano, R., Lukman, R., Lozano, F. J., Huisingh, D., & Lambrechts, W. (2013). Declarations for sustainability in higher education: becoming better leaders, through addressing the university system. *Journal of cleaner production*, 48, 10–19.

Mingazova, N.M. (2014). Modification Of the active learning methods in environmental education in Russian universities. *Procedia-Social and Behavioral Sciences*, 131, 85–89.

Sterling, S.R., & Orr, D. (2001). Sustainable education: Re-visioning learning and change (Vol. 6). *Totnes: Green Books for the Schumacher Society*.

Xiao, Z., Duritan, M.J.M., Han, D., & Zong, Y. (2024). How does Green education result in resource extraction and consumption sustainability?. *Resources Policy*, 89, 104626.

https://www.unep.org/facts-about-climate-emergency

CHAPTER 5

Challenges and Opportunities in Higher Education Internalization under NEP 2020

Dr. Ajay G Jasola* & Dr. Alok Jasola**

*Chairman, **Director

Himalayan Institute of Technology, Dehradun, Uttarakhand, India

Abstract

The National Education Policy (NEP) 2020 marks a significant milestone in India's educational framework, presenting a visionary blueprint for the internationalization of higher education. This paper, titled "Challenges and Opportunities in Higher Education Internationalization under NEP 2020," explores the obstacles and prospects inherent in this initiative. The NEP 2020 aims to elevate Indian higher education institutions (HEIs) globally by fostering international collaborations, enhancing research output, and attracting international students and faculty. Additionally, it seeks to curb the trend of Indian students studying abroad by providing competitive opportunities within India. The paper examines structural and regulatory barriers, including the complexities of India's higher education regulatory landscape, and highlights the need for a streamlined policy environment. It also addresses infrastructural challenges, emphasizing the necessity for HEIs to upgrade facilities to meet global standards and invest in faculty development programs. Opportunities presented by NEP 2020 include the potential for Indian HEIs to enhance their academic and research profiles through international collaborations and innovative approaches to internationalization, such as leveraging technology for virtual exchanges. This paper offers strategic policy recommendations to overcome identified barriers, emphasizing the need for cohesive regulatory frameworks, infrastructural investments, and capacity-building initiatives. By harnessing internationalization, India can position itself as a global leader in higher education, driving socio-economic development and contributing to the global knowledge economy.

Keywords: *National Education Policy 2020 (NEP 2020), Higher Education Internationalization, Indian Higher Education, Academic Excellence, International Collaborations, Digital Transformation*

INTRODUCTION

The National Education Policy (NEP) 2020, introduced by the Government of India, is a landmark reform aimed at transforming the educational landscape of the country. It seeks to address the various challenges faced by the Indian education system and align it with the global standards of quality and innovation. One of the key components of NEP 2020 is the internationalization of higher education, a strategic imperative designed to enhance India's global competitiveness, foster academic excellence, and drive socio-economic development. Internationalization of higher education involves integrating an international, intercultural, and global dimension into the purpose, functions, and delivery of post-secondary education. It encompasses a broad spectrum of activities, including student and faculty exchanges, international research collaborations, cross-border educational partnerships, and the development of globally relevant curricula. The NEP 2020 envisions Indian higher education institutions (HEIs) becoming global hubs of learning, attracting students and scholars from around the world, and establishing Indian universities among the top 100 globally.

The policy outlines several ambitious goals to achieve this vision. These include the establishment of international campuses by Indian universities, the creation of research-intensive HEIs that can compete on a global scale, and the development of a regulatory framework that facilitates seamless international partnerships. NEP 2020 also emphasizes the importance of fostering a diverse and inclusive academic environment that respects and celebrates cultural differences while promoting global understanding and cooperation. However, the path to internationalizing Indian higher education is fraught with challenges. The regulatory landscape governing higher education in India is complex and often fragmented, with multiple oversight bodies imposing varying standards and requirements. This can create significant barriers to international collaboration and the establishment of foreign campuses. Additionally, many Indian HEIs lack the infrastructural and institutional capacity to meet global standards, necessitating substantial investments in facilities, faculty development, and administrative capabilities.

Furthermore, the COVID-19 pandemic has introduced new challenges and accelerated the need for digital transformation in higher education. While the pandemic has underscored the importance of international cooperation and the exchange of knowledge, it has also highlighted the digital divide and the need for robust online education platforms. As such, NEP 2020's vision for internationalization must be re-evaluated in the context of these emerging realities, with a focus on leveraging technology to bridge geographical barriers and enhance accessibility. Despite these challenges, the internationalization of higher education under NEP 2020 presents numerous opportunities. By adopting best practices from leading global institutions, Indian HEIs can enhance their academic and research profiles, attract top talent, and foster an environment of innovation and excellence. International collaborations can also facilitate knowledge transfer, enhance research output, and contribute to the development of solutions to global challenges.

REVIEW OF LITERATURE

The internationalization of higher education has become a strategic imperative for countries seeking to enhance their global competitiveness, foster academic excellence, and drive socio-economic development. The National Education Policy (NEP) 2020 represents a significant shift in India's approach to higher education, emphasizing internationalization as a key component. This literature review provides an in-depth analysis of the challenges and opportunities associated with the internationalization of higher education under NEP 2020, drawing on a wide range of scholarly sources, policy documents, and case studies.

Theoretical Frameworks and Approaches

Knight (2004) defines internationalization as the process of integrating an international, intercultural, and global dimension into the purpose, functions, and delivery of post-secondary education. This broad definition encompasses a variety of activities, including student and faculty exchanges, international research collaborations, cross-border educational partnerships, and the development of globally relevant curricula. Wit (2011) further categorizes the rationales for internationalization into four key areas: academic, cultural, economic, and political, highlighting the multifaceted nature of this process.

Policy Landscape and Strategic Directions

The NEP 2020 sets ambitious goals for the internationalization of Indian higher education. According to the Government of India (2020), the policy aims to establish high-performing Indian universities among the global top 100, foster global research collaborations, and attract international students and faculty. This vision is supported by various strategic initiatives, including the establishment of international campuses, the development of research-intensive HEIs, and the creation of a conducive regulatory environment for international partnerships.

CHALLENGES IN INTERNATIONALIZATION

Regulatory Barriers

The regulatory landscape governing higher education in India is complex and often fragmented. Qamar and Samreen (2020) highlight the difficulties faced by HEIs in navigating the myriad of oversight bodies, each with its own standards and requirements. This complexity can impede the establishment of international collaborations and the setting up of foreign campuses.

Infrastructural Inadequacies

Many Indian HEIs lack the necessary infrastructural capacity to meet global standards. This includes state-of-the-art research labs, enhanced digital infrastructure for remote learning, and vibrant campus environments that appeal to international students. Agarwal (2009) underscores the urgent need for substantial investments in these areas to support internationalization efforts.

Impact of the COVID-19 Pandemic

The COVID-19 pandemic has introduced new challenges and accelerated the need for digital transformation in higher education. While the pandemic has underscored the importance of international cooperation and the exchange of knowledge, it has also highlighted the digital divide and the need for robust online education platforms. As such, NEP 2020's vision for internationalization must be re-evaluated in the context of these emerging realities, with a focus on leveraging technology to bridge geographical barriers and enhance accessibility.

OPPORTUNITIES IN INTERNATIONALIZATION

Enhancing Academic and Research Excellence

Internationalization presents significant opportunities for Indian HEIs to enhance their academic and research profiles. Marginson and Wende (2007) argue that international collaborations facilitate knowledge transfer, enhance research output, and foster innovation. Successful case studies, such as those documented by Sharma and Roy (2021), demonstrate how Indian universities have leveraged international partnerships to improve their global footprint.

Leveraging Technology for Virtual Internationalization

The COVID-19 pandemic has accelerated the need for digital transformation in higher education. Sen and Mehta (2020) discuss the potential of digital platforms to facilitate virtual exchanges and remote learning, enabling greater access to international education and collaboration. This approach can bridge geographical barriers and enhance the accessibility of global learning opportunities.

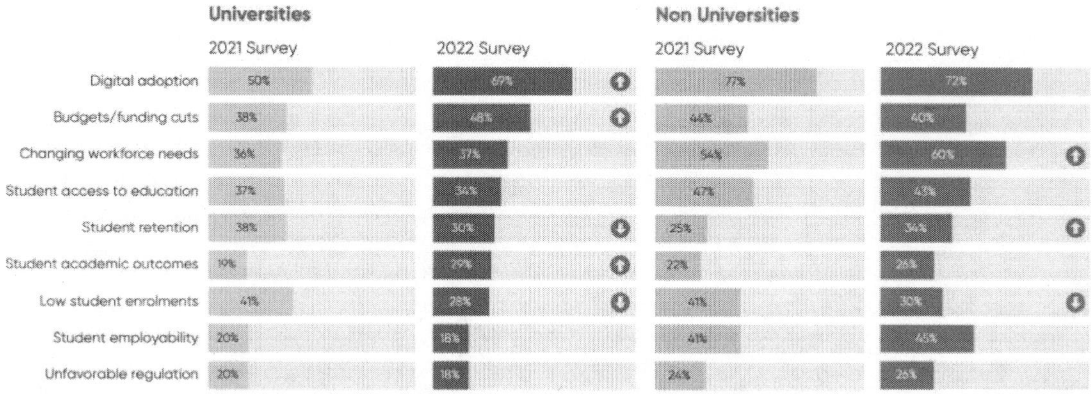

Source: Digital Transformation in Higher Education (https://www.holoniq.com/notes/2023-higher-education-digital-transformation-survey)

Comparing Student Mobility Trends

In recent years, there has been a significant increase in the number of Indian students pursuing higher education abroad. According to data from the OECD (2020), the number of Indian students enrolled in tertiary education abroad has steadily increased, making India one of the largest sources of international students globally. Countries like the United States, Canada, United Kingdom, and Australia are the most popular destinations for Indian students, attracted by the perceived higher quality of education, better infrastructure, and greater opportunities for research and employment.

Table 1: Indian Students Studying Abroad

Year	*Total Registrations*	*Year*	*Total Registrations*
2017	4,54,009	2020	2,59,655
2018	5,17,998	2021	4,44,553
2019	5,86,337	2022	7,50,365

Source: Indian Express

However, this trend has significant implications for India, both in terms of economic costs and brain drain. Indian students studying abroad contribute substantially to the economies of host countries through tuition fees and living expenses. Furthermore, the loss of talented individuals to foreign institutions represents a missed opportunity for India to harness their potential for national development.

Government Initiatives and Policy Support

The role of government initiatives in supporting internationalization efforts is crucial. Funding programs and policy incentives designed to attract international talent can significantly enhance the capacity of Indian HEIs to compete globally. According to the University Grants Commission (UGC) (2021), initiatives such as the "Study in India" program and the establishment of research parks and innovation hubs are key components of NEP 2020's internationalization strategy. These initiatives aim to create a supportive ecosystem for international education and research, fostering an environment of academic excellence and innovation

Reducing the Need to Study Abroad

The internationalization of higher education within India, as envisioned by NEP 2020, offers a pathway to reverse these trends. By enhancing the quality and global competitiveness of Indian HEIs, the policy aims to retain talented students within the country, contributing to the vision of "Atmanirbhar Bharat" (self-reliant India). If Indian universities can provide education and research opportunities at par with leading global institutions, students may no longer feel the need to go abroad for higher education.

CASE STUDY: INDIAN INSTITUTE OF TECHNOLOGY AND INTERNATIONAL COLLABORATION

Institutions like the Indian Institute of Technology (IITs) have already begun to establish themselves as global centers of excellence. Varghese (2008) notes that IITs have entered into numerous international collaborations, attracting students and faculty from around the world. These collaborations have enhanced the research capabilities of IITs and positioned them as competitive players on the global stage.

POLICY RECOMMENDATIONS

To overcome the challenges and harness the opportunities presented by NEP 2020, several strategic policy recommendations are proposed:

1. **Streamlining Regulatory Frameworks:** Simplifying and harmonizing the regulatory landscape to facilitate seamless international collaborations and the establishment of foreign campuses.
2. **Investing in Infrastructure:** Enhancing infrastructural capacity, including research labs, digital infrastructure, and campus facilities, to meet global standards.
3. **Capacity Building:** Developing faculty and administrative capabilities through targeted training programs to deliver a globalized curriculum and manage international partnerships.
4. **Promoting Digital Internationalization:** Leveraging technology to facilitate virtual exchanges, remote learning, and global research collaborations.
5. **Supporting Government Initiatives:** Encouraging the adoption of government programs and policy incentives designed to attract international talent and support internationalization efforts.

By implementing these strategic recommendations, India can position itself as a global leader in higher education, driving socio-economic development and contributing to the global knowledge economy. The successful internationalization of Indian higher education under NEP 2020 will not only enhance the quality of education but also support the broader vision of "Atmanirbhar Bharat," ensuring that the country's academic and research potential is fully realized an the global stage.

CONCLUSION

The National Education Policy (NEP) 2020 heralds a transformative era for India's higher education system, with internationalization positioned as a cornerstone for achieving global competitiveness and academic excellence. This visionary policy aims to propel Indian higher education institutions (HEIs) onto the global stage by fostering international collaborations, enhancing research output, and attracting a diverse pool of international students and faculty. While the potential benefits are immense, the path to realizing these goals is fraught with significant challenges. One of the most pressing challenges is the complex and fragmented regulatory landscape governing higher education in India. The multitude of oversight bodies, each with its own standards and requirements, creates significant barriers to the establishment of international collaborations and foreign campuses. Addressing this issue requires a concerted effort to streamline and harmonize regulations, creating a cohesive policy environment that facilitates seamless international partnerships.

Infrastructural inadequacies present another formidable challenge. Many Indian HEIs lack the state-of-the-art facilities necessary to meet global standards, including advanced research labs, digital infrastructure for remote learning, and vibrant campus environments that appeal to international students. Substantial investments are needed to upgrade these facilities, ensuring that Indian institutions can compete on a global scale. Additionally, capacity building is crucial; HEIs must invest in faculty development programs that equip educators with the skills to deliver a globalized curriculum and manage international partnerships effectively. Despite these challenges, the opportunities presented by NEP 2020 are manifold.

By embracing internationalization, Indian HEIs can enhance their academic and research profiles, attract top talent from around the world, and foster an environment of innovation and excellence. Successful case studies from institutions like the Indian Institutes of Technology (IITs) illustrate the potential benefits of international partnerships, demonstrating how global collaborations can enhance research capabilities and elevate India's position in the global educational landscape. Moreover, the COVID-19 pandemic has underscored the importance of digital transformation in higher education. Leveraging technology to facilitate virtual exchanges and remote learning can bridge geographical barriers, enabling greater access to international education and collaboration. This approach not only enhances the accessibility of global learning opportunities but also addresses the emerging realities of the digital age. To fully harness the opportunities presented by NEP 2020, several strategic policy recommendations are proposed. These include simplifying and harmonizing the regulatory framework, enhancing infrastructural capacity, developing faculty and administrative capabilities, and promoting digital internationalization. By implementing these strategies, India can position itself as a global leader in higher education, driving socio-economic development and contributing to the global knowledge economy.

The internationalization of higher education under NEP 2020 also aligns with the broader vision of "Atmanirbhar Bharat" (self-reliant India). By enhancing the quality and global competitiveness of Indian HEIs, the policy aims to retain talented students within the country, reducing the need for them to seek education abroad. This shift not only mitigates the economic costs associated with studying abroad but also harnesses the potential of Indian students and scholars for national development.

In conclusion, the successful internationalization of Indian higher education under NEP 2020 requires a nuanced understanding of the challenges and opportunities involved. By adopting a strategic approach that addresses regulatory and infrastructural barriers while leveraging the potential of digital technologies, India can realize the full potential of NEP 2020's internationalization objectives. This will not only enhance the quality of education but also support India's emergence as a global leader in higher education, driving innovation, excellence, and socio-economic development. This paper aims to contribute to the ongoing discourse on educational reform, providing a roadmap for the successful implementation of NEP 2020's internationalization vision.

REFERENCES

Agarwal, P. (2009). Indian higher education: Envisioning the future. SAGE Publications India.

Altbach, P.G., & Knight, J. (2007). The internationalization of higher education: Motivations and realities. *Journal of Studies in International Education*, 11(3-4), 290-305.

Wit, H. (2011). Trends, issues, and challenges in internationalization of higher education. Center for International Higher Education, Boston College.

Government of India (2020). National Education Policy 2020. Ministry of Human Resource Development.

Knight, J. (2004). Internationalization remodeled: Definition, approaches, and rationales. *Journal of Studies in International Education*, 8(1), 5-31.

Marginson, S., & van der Wende, M. (2007). Globalization and higher education. *OECD Education Working Papers*, No. 8.

OECD (2020). Internationalisation of higher education. Organisation for Economic Co-operation and Development.

Qamar, A., & Samreen, F. (2020). Internationalization of higher education in India: Challenges and prospects. *International Journal of Educational Management*, 34(4), 853-866.

Sen, A., & Mehta, R. (2020). Bridging the gap: Challenges and opportunities in internationalizing Indian higher education. Presented at the *Association of Indian Universities Annual Conference.*

Sharma, Y., & Roy, R. (2021). Internationalization of higher education in India: Strategies and case studies. *Indian Journal of Higher Education*, 12(2), 95-108.

Varghese, N.V. (2008). Globalization of higher education and cross-border student mobility. *International Institute for Educational Planning (IIEP)*, UNESCO.

British Council (2014). Internationalising higher education: Mapping India's global initiatives.

CHAPTER 6

Technology Integration and Digital Skills

Mitali Batra* **& Sanjeev Batra****

**Cofounder & Head L&D, **Cofounder & Client Partner,*
SANMIT Technologies, Noida, India

Abstract

In the fast-evolving landscape of technological innovation, our education policy must adapt to current needs. The National Education Policy (NEP) 2020 of India is a crucial step in this direction, marking a significant milestone in the country's educational landscape. NEP 2020 emphasizes integrating technology and developing digital skills as essential components of modern education. This paper examines various aspects of technology integration and digital skills as outlined by NEP 2020. It addresses the ethical use of technology and how to cultivate responsible digital citizenship. Additionally, it explores the role of open educational resources (OERs) and digital content repositories in bridging the digital divide and providing quality education to a broader audience. While these resources make education more accessible, digital literacy and cybersecurity education are vital to their effective use. Innovative teaching methods enabled by digital pedagogy are essential for developing these skills. By aligning with NEP 2020 principles and leveraging evidence-based strategies, educators can navigate the digital transformation of education and empower learners for success in the digital age.

Keywords: *Technology Integration, Digital Skills, Open Educational Resources*

INTRODUCTION

In the fast-evolving landscape of technological innovation, our education policy needs to be adaptable to the current needs. Need of the hour was to have an education policy which paves the way for smooth adaptation of the technology. The National Education Policy (NEP) 2020 of India is one right step in that direction and it marks an important milestone in the country's educational landscape. It has the right ingredients involving and emphasizing the integration of technology and the development of digital skills as essential components of modern education. There are multifaceted aspects of technology integration and digital skills in purview of NEP 2020's directives in education. Furthermore, there is dilemma and confusion on ethical use of technology and how one can be better digital citizen. The paper also explores the role of open educational resources (OERs) and digital content repositories in bridging the digital divide gap and accessibility of quality education to a wider audience. The availability of such educational resources and digital content repositories makes education accessible to masses but at the same time it becomes important to have digital literacy and cybersecurity education. Without having proper digital literacy and cybersecurity understanding, there is not much use of these repositories. In order to have better digital skills, there should be innovative teaching methods enabled by digital pedagogy. By aligning with the principles outlined in NEP 2020 and leveraging evidence-based strategies, educators can effectively navigate the digital transformation of education and empower learners for success in the digital age.

ETHICAL USE OF TECHNOLOGY AND DIGITAL CITIZENSHIP

The ethical use of technology and the cultivation of digital citizenship are fundamental pillars of NEP 2020, aimed at fostering responsible and ethical behaviour in the digital realm. These concepts are crucial as technology becomes increasingly integral to our daily lives and educational environments.

Ethical Use of Technology

The ethical use of technology encompasses a range of considerations, from respecting intellectual property rights to safeguarding personal data and promoting digital etiquette. Personal data safety becomes altogether much more important in today's fast changing technological landscape and one must learn how to protect their privacy and understand the implications of sharing personal information online. Promoting digital etiquette includes teaching students how to communicate respectfully and responsibly in digital spaces, understanding the impact of their words and actions on others. With widespread technology adoption, one must also need to have understanding about the environmental impact of digital devices and promoting sustainable practices. We should be encouraging the responsible disposal and recycling of electronic waste and considering the energy consumption of digital tools.

Digital Citizenship

Digital citizenship is nothing but the responsible and ethical use of technology in the society. According to Ribble and Bailey (2007), digital citizenship encompasses various aspects, including digital rights and responsibilities, digital etiquette, and digital health and wellness. Digital rights involve the right to privacy and freedom of expression. While digital responsibilities encompass that one is not engaged in or encourage cyberbullying or spreading misinformation.

Digital health and wellness focus on maintaining a healthy balance between digital and offline life, managing screen time, and preventing issues such as digital addiction and cyberbullying. It also involves understanding the physical and mental health impacts of technology use, such as eye strain and the effects of social media on mental health.

Role of Educators

Educators play a pivotal role in instilling these values in students. They do this through explicit instruction, where they directly teach concepts related to digital citizenship and ethical technology use. Modelling is another powerful tool; when educators demonstrate ethical behaviour and responsible technology use, students are more likely to emulate these practices. Experiential learning opportunities, such as project-based learning and real-world applications, allow students to practice and internalize these concepts.

Integrating Digital Citizenship into the Curriculum

Research by Hollandsworth et al. (2011) underscores the importance of integrating digital citizenship into the curriculum. This integration involves emphasizing critical thinking and ethical decision-making in online environments. For instance, students can learn to critically evaluate the credibility of online sources, understand the impact of their digital footprint, and make informed decisions about their online interactions.

Strategies for Promoting Digital Citizenship

Several strategies can help cultivate a generation of responsible digital citizens. Digital literacy programs teach students the skills they need to navigate the digital world effectively. Online safety workshops can provide practical guidance on avoiding online threats such as phishing, malware, and identity theft. Incorporating digital citizenship into the curriculum involves creating lessons and activities that address these topics. For example, students can participate in discussions about the ethical implications of emerging technologies, such as artificial intelligence and genetic engineering.

The ethical technology usage and the spreading digital citizenship are crucial to thrive in the digital age. Educators can help in developing the skills and values to navigate the digital world safely, ethically, and responsibly. NEP 2020's emphasis on these pillars reflects a commitment to creating a generation of informed, responsible, and ethical digital citizens who can contribute positively to the digital ecosystem. Through thoughtful integration of these principles into education, we can empower learners to make informed decisions and act with integrity in the digital realm.

OPEN EDUCATIONAL RESOURCES AND DIGITAL CONTENT REPOSITORIES

The National Education Policy (NEP) 2020 strongly recommends the widespread use of open educational resources (OERs) and digital content repositories to have better access to quality educational materials. OERs are freely accessible, openly licensed resources that can be used, adapted, and shared for educational purposes (Hilton, 2016). These resources play a crucial role in democratizing education by making high-quality materials available to everyone, regardless of their financial situation or geographic location. Research by Wiley and Hilton (2009) proves that OERs not only enhance student outcomes but also reduce educational costs, making education more affordable and inclusive. Open educational resources are helping in bridging the gap between people with resources and having access to premium material and people who do not have that access for any reason, be it financial strength or due to limited resources in certain educational institutions. By providing free access to a wide range of educational materials, OERs ensure that all students have the opportunity to learn from the best resources available. However, the successful implementation of OERs in education faces multifaceted challenges. Quality assurance is a significant concern, as the freely accessible nature of OERs means that there can be significant variability in the quality of the materials. Ensuring that these resources meet high educational standards is essential for their effective use. Additionally, accessibility is another critical issue. While OERs are theoretically available to everyone, students with disabilities or those without reliable internet access may still find it difficult to utilize these resources fully. Addressing these accessibility issues is crucial to maximizing the benefits of OERs. Despite the fact that OERs are openly licensed, there can be confusion over what constitutes fair use and how materials can be legally adapted and shared without violating the copyrights. Clear guidelines and robust legal frameworks are needed to overcome these complexities and ensure that educators and students can use OERs confidently and legally.

Platforms such as the National Digital Library of India (NDLI) and SWAYAM provide extensive digital resources covering various subjects and grade levels (Bhattacharya et al., 2017).

These platforms are invaluable tools for educators, offering a wealth of materials that can be used to create personalized learning experiences. By leveraging these resources, educators can design lessons that cater to the individual needs of their students, fostering a more engaging and effective learning environment. Additionally, these platforms encourage collaboration and support inquiry-based learning approaches, helping students develop critical thinking skills and a deeper understanding of the subject matter. Initiatives such as developing open textbooks and community-driven repositories can significantly improve the availability and quality of OERs and these initiatives should be encouraged. Open textbooks offer a cost-effective alternative to traditional textbooks and at the same time allows educators to customize content to better meet their students' needs. Community-driven repositories, where educators and experts can contribute and review materials, ensure a continuous improvement of the quality and relevance of the resources available. These initiatives promote a culture of sharing and collaboration, essential for the sustainable growth of the OER movement. Having focus on the widespread adoption and effective implementation of OERs and digital content repositories, NEP 2020 aims to create an equitable educational landscape where quality resources are accessible to all learners. This approach not only supports the academic success of students but also fosters a more inclusive and collaborative educational environment. With the collaborative efforts of educators, policymakers, and the community, the full potential of OERs can be realized which will help a better future for all learners.

DIGITAL PEDAGOGY AND INNOVATIVE TEACHING METHODS

The National Education Policy (NEP) 2020 emphasizes the critical need to embrace digital pedagogy and innovative teaching methods to enhance student engagement and improve learning outcomes. Digital pedagogy refers to the use of digital tools and resources to facilitate and enhance the teaching and learning process (Johnson et al., 2015). This approach recognizes the transformative potential of technology in education, advocating for methods that make learning more interactive, personalized, and relevant to real-world contexts. One of the innovative methods highlighted by NEP 2020 is the flipped classroom model. In this approach, traditional lecture and homework elements are reversed. Students engage with instructional content, often delivered via video, outside of class. In-class time is then devoted to exercises, projects, or discussions that deepen understanding and allow for personalized teacher support. This model promotes active learning and enables students to apply concepts in practical settings, thereby enhancing comprehension and retention.

Gamification is another innovative strategy recommended by NEP 2020. By incorporating game elements such as points, badges, and leaderboards into educational activities, gamification can increase student motivation and engagement. This approach makes learning more enjoyable and can cater to various learning styles, encouraging students to participate actively and persistently. Hybrid or blended learning, which combines traditional face-to-face instruction with online learning, is also emphasized. This method offers flexibility and personalized learning experiences, allowing students to progress at their own pace. Blended learning environments can provide a rich array of resources and activities, making learning more engaging and accessible. Research by Kumar and Vigil (2019) highlights the effectiveness of educational apps, virtual reality (VR) simulations, and online collaborative platforms in creating immersive learning experiences tailored to diverse learning styles. Educational apps can provide interactive and personalized learning experiences, making complex subjects more understandable and engaging. VR simulations offer immersive experiences that can bring abstract concepts to life and provide hands-on learning opportunities in a safe environment. Online collaborative platforms enable students to work together on projects, share ideas, and receive feedback, fostering a sense of community and collaboration.

Educators can leverage these digital tools to design authentic learning tasks that encourage creativity, critical thinking, and problem-solving. By integrating technology into their teaching practices, educators can create dynamic and interactive learning environments that engage students and enhance their learning experiences. Authentic learning tasks, which are relevant to real-world scenarios, help students develop practical skills and apply their knowledge in meaningful ways. However, the successful integration of digital pedagogy requires more than just access to technology. Professional development programs and ongoing support are essential to equip educators with the skills and competencies needed to use technology effectively in their teaching practices (Ertmer et al., 2012). These programs should focus on training educators to design and implement digital learning activities, manage digital classrooms, and use data to inform instruction. Ongoing support, including access to technical assistance and professional learning communities, is crucial for helping educators stay updated with the latest technological advancements and pedagogical strategies.

NEP 2020 advocates for the adoption of innovative teaching methods enabled by digital pedagogy to enhance student engagement and learning outcomes. By leveraging digital tools such as educational apps, virtual reality simulations, and online collaborative platforms, educators can create immersive learning experiences that foster 21st-century skills, including critical thinking, creativity, collaboration, and communication (Kumar & Vigil, 2019). These skills are essential for success in the modern, interconnected world. NEP 2020's insistence on digital pedagogy and innovative teaching methods reflects a forward-thinking approach to education. By integrating technology into teaching practices and adopting innovative strategies, educators can enhance student engagement and learning outcomes. Through these efforts, NEP 2020 aims to create a more dynamic, inclusive, and effective educational system that prepares students for the challenges and opportunities of the 21st century.

DIGITAL LITERACY AND CYBER SAFETY EDUCATION

The National Education Policy (NEP) 2020 recognizes the critical importance of digital literacy and cybersecurity education in preparing students for the digital age. Digital literacy skills are necessary for students to navigate, understand, and create information in the digital world. These skills include information literacy, media literacy, and online safety (Livingstone et al., 2017). Information literacy involves the ability to locate, evaluate, and use information effectively. Media literacy focuses on critically analysing media content and understanding its effects. Online safety includes understanding how to protect personal information and navigate the internet responsibly. Similarly, cybersecurity education is crucial for raising awareness about online threats, data privacy, and responsible digital behaviour (Furnell & Dowland,

2017). Cybersecurity education teaches students about potential online risks, such as hacking, phishing, and identity theft, and how to protect themselves and their data from these threats. This education is not just about technical skills but also about fostering a culture of responsible and ethical behaviour in the digital realm.

Integrating digital literacy and cybersecurity education into the curriculum is essential. This integration ensures that students are not only users of technology but also understand the ethical implications and responsibilities that come with it. By incorporating these topics into the curriculum, educators can empower students to navigate the digital world safely, ethically, and responsibly. This education prepares students to be informed digital citizens who can critically evaluate digital content, understand the importance of data privacy, and behave responsibly online. Research-based interventions enhances awareness of cybersecurity risks. Interactive workshops provide hands-on experiences where students can practice identifying and mitigating cybersecurity threats. Scenario-based learning activities simulate real-world situations, helping students understand the consequences of their actions and make informed decisions. Real-world case studies offer insights into actual incidents of cyber threats and how they were managed, providing practical lessons on cybersecurity. Collaboration with parents, community stakeholders, and industry partners is essential to create a culture of digital responsibility and resilience across all facets of society. The National Education Policy (NEP) 2020 emphasizes the integration of digital literacy and cybersecurity education at all educational levels. This integration aims to equip students with the knowledge and skills to navigate the digital landscape safely and responsibly. Implementing these initiatives requires collaboration among policymakers, educators, parents, and industry stakeholders to ensure comprehensive digital literacy and cybersecurity education for all learners. Policymakers need to provide the necessary framework and resources to support these educational initiatives. Educators require professional development to effectively teach digital literacy and cybersecurity. Parents and community members must be engaged in reinforcing these lessons outside of the classroom. Industry partners can contribute by providing up-to-date information on emerging threats and best practices.

CONCLUSION

NEP 2020 represents a visionary blueprint for transforming education in India, emphasizing technology integration and digital skills development as key elements of educational reform. By embracing NEP 2020's principles and leveraging evidence-based strategies, educators can successfully navigate the digital transformation of education and empower learners for success in the digital age. Collaboration among stakeholders, investment in professional development, and ongoing research are essential to fully realize the potential of technology integration and digital skills development in education, thereby shaping a brighter future for learners nationwide. Through concerted efforts and a commitment to excellence, educators can harness the power of technology to create inclusive, equitable, and empowering learning environments that prepare students to thrive in the complex and interconnected world of the 21st century. Technology integration and digital skills are integral to modern education, enhancing teaching and learning experiences, promoting digital citizenship, and preparing students for the digital future. By prioritizing ethical technology use, leveraging open educational resources, embracing innovative teaching methods, and implementing NEP 2020 initiatives, educators can foster a digitally literate and ethically conscious generation. Collaboration among stakeholders and ongoing investment in professional development are key to maximizing the potential of technology integration in education and shaping a brighter future for learners worldwide.

REFERENCES

Ertmer, P.A., Ottenbreit-Leftwich, A.T., Sadik, O., Sendurur, E., & Sendurur, P. (2012). Teacher Beliefs and Technology Integration Practices: A Critical Relationship. *Computers & Education*, 59(2), 423-435. doi:10.1016/j.compedu.2012.02.001.

Furnell, S., & Dowland, P. (2010). A Conceptual Framework for Supporting Security Education and Awareness. *Information Management & Computer Security*, 18(1), 26-38. doi:10.1108/09685221011035235.

Hilton, J. (2016). Open Educational Resources and College Textbook Choices: A Review of Research on Efficacy and Perceptions. *Educational Technology Research and Development*, 64(4), 573-590. doi:10.1007/s11423-016-9434-9.

Hollandsworth, R., Dowdy, L., & Donovan, J. (2011). Digital Citizenship in K-12: It Takes a Village. *TechTrends*, 55(4), 37-47. doi:10.1007/s11528-011-0510-z.

Johnson, L., Adams Becker, S., Estrada, V., & Freeman, A. (2015). NMC Horizon Report: 2015 Higher Education Edition. The New Media Consortium.

Kumar, S., & Vigil, K. (2019). Educational Technology: Effective Use and Practices. *Journal of Educational Technology Systems*, 47(1), 13-27. doi:10.1177/0047239518793768.

Livingstone, S., Van Couvering, E., & Thumim, N. (2008). Converging Traditions of Research on Media and Information Literacies: Disciplinary, Critical, and Methodological Issues. *Handbook of Research on New Literacies*, 103-132.

Ribble, M., & Bailey, G. (2007). Digital Citizenship: Using Technology Appropriately. *Learning & Leading with Technology*, 35(1), 12-15.

CHAPTER 7

Exploring the Role of Agribusiness Incubators on Agricultural Innovation and Entrepreneurship

DISHA GULIA[*] & DR. SAPNA ARORA[**]

[*]Research Scholar, [**]Associate Professor

Department of FBMED, National Institute of Food Technology Entrepreneurship and Management, India

ABSTRACT

The transformation of Indian agriculture into an appealing and lucrative business requires a move from agriculture to agribusiness. In light of this, the government has established business incubators facilities to encourage entrepreneurship in agro food sector. These agribusiness incubators will open up the incubation landscape to the micro section of the huge rural economy. They support small business development initiatives, promote growth within local economies, and provide as a forum for technology transfer. The present paper highlights the role of agribusiness incubators in fostering technological innovations in the agro food sector and to explore the agribusiness incubation scenario in India.

Keywords: *Agribusiness Incubators (ABI), Entrepreneurship, Agriculture*

INTRODUCTION

Business Incubators have proliferated as tools for entrepreneurship development over the past few decades. Business incubators, on the other hand, are a relatively new but quickly expanding concept for India's entrepreneurial environment. Business incubators are institutions that promote early-stage companies by offering technology and business infrastructure, other support services, and internal and external networks. (Peters et al., 2004). The National Business Incubation Association (NBIA, 2014) addressed business incubation as "a business support process that accelerates the successful development of start-up and fledgling companies by providing entrepreneurs with an array of targeted resources and services." The incubator offers a range of value added services to boost the chances of survival and to increase the profitability of incubatees. Every incubator has their own strategy and incubation method in accordance with its establishment's objectives as well as the needs of client firms but the universal components of any incubator stay the same, nevertheless. Despite the significance of incubators in linking entrepreneurship to economic development, their objectives, methods, and effects on developmental outcomes in underdeveloped nations are not well known.

REVIEW OF LITERATURE

In 1958, Joseph Mancuso—best recognised as the inventor of incubators—started the first business incubator at a sizable, abandoned plant in New York's Batavia Industrial Center (NBIA, 2007). According to Haugen (1990), business incubators were created as a tool for economic growth during the post-industrial era in order to occupy vacant premises and lower the unemployment rate. Such worries have increased as a result of suburbanization and the

Table 1: Incubation and Incubators – Global Scenario

Country	China	USA	UK	India
No. of Incubators	3000+	1500+	400+	520+
Type of Incubators	• Government supported • Corporate • Academic	• Startup incubators • For profit • Non-profit • Govt. supported	• Independent • Academic • Corporate • Govt. supported	• Corporate • Independent • Academic • Govt. supported
Global Innovation Index Rank (2021)	12	3	4	46
Government Objectives of Incubation Programs	Guidance for developing high tech industries	Fostering innovation and commercialization of research	Research, technological development & innovation (RTDI)	Innovation, Employment generation & solving specific local problems
No. of Unicorn Startups	301	487	39	81

Source: NASSCOM Report 2018, Hurun Global Unicorn List 2021, Global best practices report on incubation and acceleration unitus VC.

concurrent large-scale industrial emigration from metropolitan cores. As a result, the purpose of business incubators is to stabilize the local economy (Wiggins and Gibson, 2003). In the 1950s, the first BIs were founded in the USA (Adkins, 2002). The idea gained popularity in the 1980s and spread to the rest of the world in a number of ways (EC, 2002). Business Incubators can be clubbed in three categories: first generation, second generation, and third-generation. The first generation of business incubators existed between 1959 and 1979; the second generation ran from 1980 to 1999; and the third generation started in 2000 and lasted until 2012 (Shepard, 2013). These first-generation BIs (Barrow, 2001) provided accessible office space and pooled resources. Infrastructure is the fundamental purpose shared by all types of business and the basis of their service offerings (Allen and McCluskey, 1990). It comprises of shared office space that incubators rent under advantageous terms (Bergek and Norrman, 2008). Additionally, Business incubators have small production facilities that are accessible to client firms. However, the second generation incubator is prepared to offer more sophisticated and practical services, such as business planning, marketing, mentoring, and access to funding. The third generation is credited with enhanced incubator benefits through networking services, emphasizing preferred access to technology partners, prospective clients, product manufacturers, and investors (Apa et al., 2017).

The results of the historical analysis show that the fundamental idea behind business incubators has not altered over time. The concept of guiding a young, budding firm until it can function independently, similar to what an incubator does for young premature babies and even today, the inspiration for business incubators comes from newborns. However, over time, the characteristics of business have changed and become more sophisticated and aligned to contextual and situational forces. The three generations of incubatees in business incubators have extremely varied characteristics. Unlike first or the second generation incubatees, who

Table 2: Different Generations of Incubators

Generations of Incubators	*Attributes/Offerings*
First Generation (Economies of scale)	– Physical Infrastructure • Office Space • Conference Rooms • General office Equipments – Shared Facilities • Security • Computer • Receptionist Services
Second Generation	– Business Assistance • Access to Finance • Business Planning • Coaching • Accounting & Legal services • Market Advice • Sales & Marketing • Business Planning – Technical Assistance • Research & Development • Technology Transfer • Technical skills • Intellectual Property
Third Generation	– Access to external coaching, knowledge & legal Services – Access to industry experts & mentors – Access to networking • Customers • Suppliers • Peers • Angel Investors • Strategic partners

are often older in the business, third generation tenants are new enterprises started by serial entrepreneurs.

Agribusiness Incubators

Agribusiness incubation is a strategy that emphases on fostering creative, start-up stage firms with huge potential for growth to become successful agribusinesses by providing services, enhancing the value of, or establishing connections with farm producers (Hjortso et al., 2017). The present study's contribution is to expand on the body of literature by thoroughly understanding the agribusiness incubation scenario in India. This study will aim to answer the following significant research questions:

- RQ1: To study the agribusiness incubation landscape in India

- RQ2: To study the role of agribusiness incubators in fostering innovations in agro food sector

AGRIBUSINESS ECOSYSTEM IN INDIA

Considering the crucial role that agriculture plays in local economies and the importance of smallholder farmers, supporting agriculture-related entrepreneurship with a focus on the small producers will assist the industry adapt to the quickly shifting global environment. However, for such transformation to take place in the industry, there must be enabling conditions that foster an entrepreneurial ecosystem and encourage farmers to take risks and run agribusiness start-ups while also allowing other sector stakeholders to play a larger and more effective role that will accelerate the sector's growth.

The agro food sector in India faces numerous difficulties, including technological and market risks, natural disasters, price volatility, infrastructure gaps, and an absence of access to capital. To combat these challenges, business incubators offer wide variety of specialized services to the agro food SMEs.

Types of Agribusiness Incubators

There are three main categories of agribusiness incubators as recognized by the World Bank's InfoDev programme.

1. **Agribusiness value chain and sector development incubators:** These incubators are experts at connecting crucial value chain components, giving market access, and assisting in the development of new businesses that close value chain gaps and aims to strengthen the entire agriculture sector and allied fields.
2. **Agribusiness research and commercialization incubators:** These incubators facilitate the technology transfer between research centers and higher education institutions and encourage development of new ventures by commercialization of new technology.
3. **Technology transfer incubators:** These incubators concentrate on promoting the technology transfer at either the high-tech ends such as or low-tech ends. The low-tech ends include innovation at the under development rural level and the high-tech ends are related to collaborative technology ventures, venture capital and intellectual property rights.

Small groups of aspiring entrepreneurs are identified and mobilized by these agribusiness incubators, who then support their development by offering a variety of services which includes shared office facilities and useful equipment, strategy development, technology, financing, coaching, and access to external networks.

AGRIBUSINESS INCUBATION LANDSCAPE IN INDIA

Incubators are getting attention all around the world for creating new businesses in research institutes and universities in the areas of technology, service, business ideas, etc. Through its revitalized flagship programmes, the Government of India has also made an effort to revive the ecosystem for technology and business development in a number of areas. Small business owners suffer with a lack of resources, technologies, knowledge, and market data. Farmers may have more chances to sell their goods and technology on a worldwide scale by leveraging and combining programmes and services to boost agribusinesses and increasing collaborations through entrepreneurial development, innovation, and quality enhancement.

The development of agribusiness incubation is still in its infancy in India's agriculture sector. Though many agribusiness incubators have been established since the first one by ICRISAT was founded in India in 2003, their success rate has not been particularly high. Therefore, it is intended for the current agribusiness incubators to be improved as needed and for new ones to be established through the provision of grants-in-aid under the scheme. Annexure 1 given below lists the all the agri business incubators established under different institutions in India.

Annexure 1: List of Agri Business Incubators in India

Affiliation	*Name of Incubation Centre*	*State*
RKVY-RAFTAAR – KPs (5)	MANAGE-Centre for Innovation and Agripreneurship, Hyderabad	Andhra Pradesh
	AAU-North East Agriculture Technology Entrepreneurs Hub, Jorhat	Assam
	ICAR-IARI Pusa Krishi, Delhi	New Delhi
	CCSNIAM ABI, Jaipur	Rajasthan
	UAS-Krishik ABI, Dharwad	Karnataka
RKVY-RAFTAAR – (R-ABI) (24)	CCSHAU R-ABI, Hisar	Haryana
	UTKARSH PDKV, Akola	Maharashtra
	Central Institute for Research on Cotton Technology, Mumbai	Maharashtra
	IGKV R-ABI, Raipur	Chattisgarh
	IVRI R-ABI Izzatnagar	Uttar Pradesh
	Jawahar R-ABI, Jabalpur	Madhya Pradesh
	Nutrihub, Hyderabad	Andhra Pradesh
	SKN Agri Business Incubator, Jaipur	Rajasthan
	KAU R-ABI, Thrissur	Kerala
	IIT-BHU R-ABI, Varanasi	Uttar Pradesh
	IIM-KSP FIED, Kashipur	Uttarakhand
	ICAR-NIVEDI NaaViC, Bengaluru	Karnataka
	BAU-Sabour R-ABI, Sabour	Bihar
	Anand Agricultural University R-ABI, Anand	Gujarat
	Punjab Agri Business Incubator, Ludhiana	Punjab
	ABIS TBI, Coimbatore	Tamil Nadu
	ANGRAU Poshan Incubator, Tirupati	Andhra Pradesh
	IIT-KGP AFBIC, Kharagpur	West Bengal

Affiliation	*Name of Incubation Centre*	*State*
	ICAR-NRRI Vikas R-ABI, Cuttack	Odisha
	SKUAST R-ABI, Jammu	J & K
	Him Palam R-ABI, Palampur	Himachal Pradesh
	College of Horticulture and Forestry R-ABI, Pasighat	Arunachal Pradesh
	COFCAU R-ABI, Lembucherra	Tripura
	Mizoram R-ABI, Aizawl	Mizoram
ICAR Incubation Centres (13)	IARI, ZTM-BPD, Delhi	New Delhi
	A IDEA-NAARM, Hyderabad	Andhra Pradesh
	ICAR-NINFET, Kolkata	West Bengal
	National Dairy Research Institute TBI, Karnal	Haryana
	National Research Centre on Pig, Guwahati	Assam
	Central Institute Of Freshwater Aquaculture, Bhubaneswar	Odisha
	Central Institute of Fisheries Technology ABI, Cochin	Kerala
	National Research Center on Meat, Hyderabad	Andhra Pradesh
	Central Institute of Brackish Water Aquaculture, Chennai	Tamil Nadu
	Indian Institute of Horticultural Research, Bengaluru	Karnataka
	Central Plantation Crops Research Institute, Kasaragod	Kerala
	Indian Institute of Spices Research, Calicut	Kozikhode
	Central Institute for Subtropical Horticulture, Lucknow	Uttar Pradesh
IIM Incubation Centres (5)	IIM-B NSRCEL, Bengaluru	Karnataka
	IIM-A CIIE.CO, Ahmedabad	Gujarat
	IIM-C Innovation Park, Kolkata	West Bengal
	IIM-L Incubator, Lucknow	Uttar Pradesh
	IIM-U Incubation Center, Udaipur	Rajasthan
IIT/NIT Incubation Centres (4)	IIT-BHU MCIIE, Varanasi	Uttar Pradesh
	NIT-C TBI, Calicut	Kerala
	IIT-Kanpur SIIC, Kanpur	Uttar Pradesh
	Foundation for Technology and Business Incubation, Rourkela	Odisha
University/College Incubation Centres (13)	Manipal University Technology Business Incubator, Manipal	Karnataka
	Ahmedabad University Support Foundation, Venture Studio, Ahmedabad	Gujarat
	SSCBS Innovation and Incubation Foundation, Delhi	New Delhi
	Sathyabama TBI, Chennai	Tamil Nadu
	Jain University Incubation Center, Bengaluru	Karnataka
	Business Innovation & Incubation Center, Kottayam	Kerala

Affiliation	*Name of Incubation Centre*	*State*
	NEC-BIC, Kovilpatti	Tamil Nadu
	Amrita TBI, Kollam	Kerala
	JSS, SJCE-STEP, Mysuru	Karnataka
	KLE-CTIE, Hubli	West Bengal
	Mizoram University Incubation Center, Aizwal	Mizoram
	BITS-PIEDS, Pilani	Rajasthan
	SVCETEDU TBI, Tirupathi	Andhra Pradesh
Central/State Govt./NIDHI/ DST/MSME Incubation Centres (20)	Institute for Industrial Development, Lucknow	Uttar Pradesh
	Scitech Park, Pune	Maharashtra
	ICRISAT-AIP, Hyderabad	Andhra Pradesh
	Kerala Startup Mission, Trivandrum	Kerala
	Kerala Agro Park, Ernakulam	Kerala
	AIC-RNTU, Raisen	Madhya Pradesh
	MITCON TBIC, Pune	Maharashtra
	COE-AGRI, Bengaluru	Karnataka
	Forum for Innovation Incubation Research & Entrepreneurship, Margao	Goa
	Centre for Advancing and Launching Enterprises, Gandhinagar	Gujarat
	NIF Incubation and Entrepreneurship Council, Gandhinagar	Gujarat
	Bannari Amman Institute of Technology TBI, Sathyamangalam	Tamil Nadu
	DLabs Incubator Association, Hyderabad	Andhra Pradesh
	G H Raisoni TBI Foundation, Nagpur	Maharashtra
	iCreate, Ahmedabad	Gujarat
	Global Incubation Services, Bengaluru	Karnataka
	AIC-NITTE, Mangalore	Karnataka
	IIT-P Foundation for Innovators in Science and Technology, Patna	Bihar
	Startup Oasis, Jaipur	Rajasthan
	IIIT-CIE, Hyderabad	Andhra Pradesh
Non-Profit/ IKP Incubation Centres (13)	Indigram Labs Foundation, Delhi	New Delhi
	Startup Oasis, Jaipur	Rajasthan
	KIIT Business Incubator, Bhubaneswar	Odisha
	Villgro Innovations Foundation, Chennai	Tamil Nadu
	Jammu Kashmir Startups Association, Jammu	J & K

Affiliation	*Name of Incubation Centre*	*State*
	T-Hub, Hyderabad	Andhra Pradesh
	LSI-IKP, Hyderabad	Andhra Pradesh
	EDII-PHBIF, Kamatchipuram	Tamil Nadu
	NASSCOM-ERNET, Bengaluru	Karnataka
	SRIX, Warangal	Telangana
	Centre for Entrepreneurship Development, Hyderabad	Andhra Pradesh
	UnLtd, Mumbai	Maharashtra
	MAGIC-CMIA, Aurangabad	Maharashtra
Private Incubation Centres (2)	iB Hubs, Gurgaon	Haryana
	Gomassive Incubator Pvt. Ltd., Noida	Uttar Pradesh

ABI-ICRISAT

The Agri-Business Incubation (ABI) programme is a project of ICRISAT under the Agribusiness and Innovation Platform (AIP), which was established in 2003 in collaboration with the Department of Science and Technology (DST), central Government of India, with the goal of fostering developments and agribusiness endeavours through public-private partnerships. The ABI programme continues to be a trailblazing initiative in the country's agricultural sector and seeks to maximise the success rate of start-up businesses by providing the finest chances with less risk. The main objective of ABI is to support agri-innovations in the industry and to speed up the commercialization of agro technology by encouraging and assisting agribusiness projects.

Startup Agri-Business Incubation Program (SAIP) under RKVY-RAFTAAR

The Ministry of Agriculture and Farmers' Welfare in colloboration with the Government of India has launched an important programme called the Rashtriya Krishi Vikas Yojna (RKVY) that aims to improve infrastructure facilities in agriculture and other related fields to financially facilitate the incubation environment and to encourage agripreneurship and agribusiness. A financial assistance programme called the Startup Agri-Business Incubation Program will be given to prospective startups that have come up with a product based on novel approaches, procedures, goods, services, and business models in the agro food sector. The programme will award up to ₹25 lakhs to each startup. The programme selection committee's decision in this matter will be final. The applicant incubatees will be given funding in accordance with their true needs and the committee's evaluation of their business ideas.

RKVY-RAFTAAR Agribusiness Incubators (R-ABIs)

This plan calls for the creation of new agribusiness incubators as well as the strengthening of current agribusiness incubators as R-ABIs by providing them with the infrastructure, tools, and personnel they require. Agripreneurs at various stages of the business life cycle (idea generation/ growth) will be invited to apply to these R-ABIs, giving them the chance to create innovations in the agricultural and related industries. These innovations include, for instance, technology approaches, procedures, goods, services, and business models that could boost efficiency or productivity in agriculture and related activities, as well as raise farmer compensation, the rural economy, and wealth creation.

Business Planning and Development (BPD) Units under ICAR

The Indian Council of Agricultural Research (ICAR) launched the BPD Units in 2008 in an effort to spur reforms in the Council's research and development practices. By collaborating with entrepreneurs, start-up businesses, and both public and private sector companies, BPD Units were established by the NAIP with the goal of encouraging and supporting technologists and scientists as they transform their innovative research ideas into sound commercial ventures. In addition to office space and access to information and communication technology, BPD is anticipated to offer variety of services, including advice on capacity building, legal issues, marketing, technical, IPR, and financial matters. BPD is thought to be an effective strategy for promoting the expansion of sustainable business endeavours.

Figure 1: Geographical Spread of BPD Units

Source: Agri-Business Incubation (ABI) Program, ICRISAT.

DISCUSSIONS

The agribusiness incubators have built effective and well-defined incubation processes to draw incubatees and institutional funding. They are managed by experts and maintain a strong focus on their objectives. Additionally, these agribusiness incubators provide a variety of services that enable startups to grow. These include access to open markets and quality infrastructure, as well as a well-established ecosystem of early-stage investors. The services provided by incubators increased the probability of startup agribusinesses surviving their first three years of operation. Agribusiness incubators are evolving more specialised support models as they get older. Some, like Villgro, have expanded beyond the agro industry; others, like ABI-ICRISAT, have started imparting their methodology to other nations; and still others, like ABI-ICRISAT and Villgro, now have started offering technological support to different incubators. Some Incubators often play the role of fund managers as well.

Services Offered of ABI

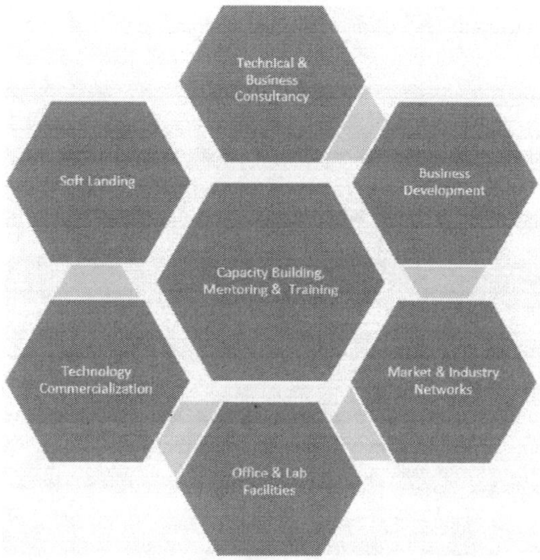

Typically, incubator programmes offer coworking spaces and startup money, but the success of these programmes depends on the level of their training and guidance. For instance, the mentors at Villgro have an average of 20 years of expertise and come from a variety of fields. (Villgro Annual Report, 2018)

A dynamic and adaptable model, such as a business incubator, is needed to support farmers in the altered agricultural development scenario, which calls for innovation and feedback mechanisms for its growth. This will help farmers enhance the efficiency of the sector and, consequently, their standard of living. The incubators provide research institutions the much-needed viewpoint of market-oriented research along with commercialization of technology. The incubators assist local farmers and budding entrepreneurs by providing them the chance to move beyond farming and engage in more significant mechanisms of economic contribution through the formation of new enterprises and job prospects. By facilitating investments (financial and physical) into the value chain, facilitating farmer development, providing researchers with

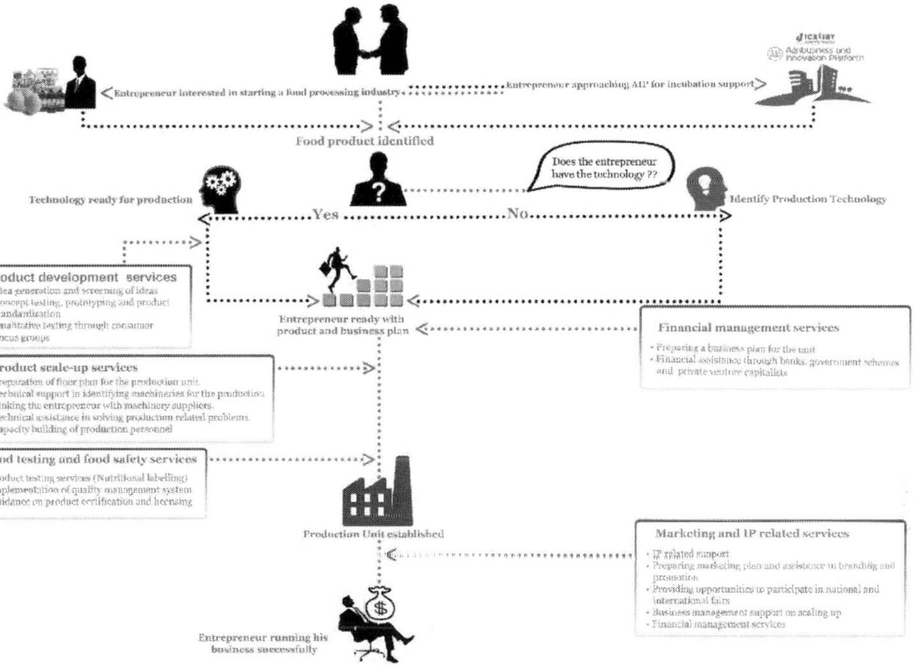

Figure 2: Specilized Incubator Services for Agro Food SMEs
Source: https://oar.icrisat.org/7259

a channel for feedback, and facilitating the growth of SMEs that improve and increase rural non-farm employment opportunities, agribusiness incubators not only become a channel for commercializing technologies but also add value to primary agricultural product and support the development of livelihoods.

Funding

While all of the incubators examined in this research have stimulated large investments in the agricultural sector, it is vital to keep in mind that they are all heavily dependent on donations and government money to operate. This is due to the fact that they concentrate on assisting high-risk startups and, in the majority of cases, smallholder farmers who are unable to pay for all of their services.

Networks

Agribusiness incubators must have strong relationships with early-stage commercial investors and should be given additional incentives. This might take a variety of shapes, like investors serving on selection of clients or acting as mentors to these client firms. Different forms of cooperation that involve investors becoming sponsors, making investments in incubators, or giving incubators financial rewards for achieving particular milestones can also be considered. For creating pace for entrepreneurship and innovation in agriculture, NIABI was designed as a global network specifically for agri-business incubators and agri-business startups. The NIABI, managed by ABI-ICRISAT aids agripreneurs in connecting with other aspiring entrepreneurs across nation and in promotion of technologies among states. The Global Agri-Business Incubation (GABI) Network, created by ABI-ICRISAT to further improve operations, will give the incubatees additional networking opportunities and access to new markets.

Risk Mitigation

Technology transfer incubators focus on a variety of technologies and industries, in contrast to agricultural value chain incubators. Incubators successfully implement risk diversification plans that reduces exclusive exposure to a sector or a firm by varying their technological and coaching support. They also do so by providing seed investments in agro start-ups firms that deal in multiple areas and diversified products. A more concentrated set of hazards are present for agricultural value chain incubators because they are often focused on fostering businesses across a small number of commodities. Operational risks associated with logistics and customer negotiations are present in agricultural value chain incubators, but these risks can be minimized by selecting the appropriate agricultural supply chains and structuring the incubation process.

Promoting Innovations in Agro Food Sector

Agribusiness incubators can help, for instance, with the creation of competitively strong agribusiness environments where industry executives typically agree that the best course of action is to become more and more knowledgeable about a smaller and smaller sector or market domain. Incubators help agripreneurs by doing market research, testing new products, and working on commercial demonstration of products. Early-stage agribusiness start-ups can more easily and swiftly adopt new technology with the aid of incubators. They can help with the creation of value chain structures that cater to more specialized market sectors. Incubators play quite important role in sector level learning of agro food sector.

Impact of Agribusiness Incubators

The advantages of an incubator differ for the many ecosystem players. while promoting corporate growth and the key benefits continue to be the development of jobs, grassroots innovators have the opportunity to scale their concepts for commercialization, universities and R&D facilities receive a means of commercialization access to their technologies for businesses innovative products and ideas, achieving financing agencies benefit from economies of scale and possibility to fund potential projects at a lesser cost transaction costs. These incubators aids in the economic development of the nation by creating jobs, Tax income growth and the expansion of various industries.

Value Chain Selection

The issues that value chain incubators and technology transfer incubators encounter are distinct because they use different business models. The decision of which value chain to concentrate on is one of the most significant problems that agricultural value chain incubators encounter at the outset.

CONCLUSION

Entrepreneurship and agribusiness initiatives need to be supported in order for an ecosystem connecting farmers with markets to thrive. Such endeavors will stimulate the local economy and interact with farmers more effectively than other initiatives, enabling them to enter the value chain. It will also make it possible to adapt agricultural technologies from research facilities to regional conditions. Agribusiness incubators greatly aids the development of the novel technology products and value added supply chains that enable small farmers and agricultural firms to prosper in emerging nations and rising markets. They play a significant role in these economies, increasing affluence and poverty alleviation. Therefore, there is growing interest in promoting policies and schemes that speed up or incubate agribusiness development. However, considering the demand and supply of innovations in agriculture sector, there is an urgent need to increase the number of agri-business incubators currently in place and further improve the environment for encouraging agri-business by facilitating funding, better support networks, training programmes of entrepreneurs in agri-business, and adoption of new models of agri-business incubators.

REFERENCES

Apa, R., Grandinetti, R., & Sedita, S. R. (2017). The social and business dimensions of a networked business incubator: the case of H-Farm. *Journal of Small Business and Enterprise Development.*

Adkins, D. (2002). A Brief History of Business Incubation in the United States. National Business Incubation Association, Athens, Ohio.

Allen, D.N., McCluskey, R. (1990). Structure, policy, services, and performance in the business incubator industry. *Entrepreneurship: Theory & Practice* 15 (2), 61–77.

Barrow, C. (2001). Incubator: A Realist's Guide to the World's New Business Accelerators. John Wiley & Sons Ltd., West Sussex, UK.

Bergek, A., Norrman, C. (2008). Incubator best practice: A framework. *Technovation* 28 (1–2), 20–28.

EC (2002). Benchmarking of Business Incubators, Final Report. Brussels.

Haugen, T.M. (1990), 'Getting a Head Start: The Rise of Business Incubators,' *Business Forum* 15, (1).

Hjortsø, C.N.P., Alexander, I.K., and Hernandez Chea, R.R. (2017). Experiences and Lessons Learned from the UniBRAIN Agribusiness Incubation Programme. *IFRO Report, No. 266.* Department of Food and Resource Economics, University of Copenhagen

NBIA (2007). The history of incubation, Available at: http://www.nbia.org/resource_center/what_is/beginnings_of_inc/index.php

Peters, L., Rice, M., & Sundararajan, M. (2004). The role of incubators in the entrepreneurial process. *The Journal of Technology Transfer*, 29(1), 83–91

NBIA (2009), "What Is Business Incubation?" Available at: http://www.nbia.org/resource_library/what_is/

Sharma KK, Karuppanchetty SM, Aravazhi S, Philroy Jonathan, Divya Nancy and Sharma Sangya (2013). A Decade of ABI Happenings. International Crops Research Institute for the Semi-Arid Tropics, Patancheru, Andhra Pradesh, India, p. 268.

Shepard, J. M. (2013). Small Business Incubators in the USA: A Historical Review and Preliminary Research Findings. *Journal of Knowledge Based Innovation in China*, 5(3), 213–233.

Villgro's annual report 2018/2019, p. 80.

Wiggins, J., and Gibson, D. V. (2003). Overview of US Incubators and the Case of the Austin Technology Incubator. *International Journal of Entrepreneurship and Innovation Management*, 3(1-2), 56-66.

CHAPTER 8

Green Synthesised Nanoparticles for Sustainable Agriculture: Optimising Growth, Protection, and Productivity of Crop, Aquaculture and Studying its Impact on Employment in Rural India

Sakshi Kumari*, Varun Ghai* & Ayush Tiwari**

*Scholars, St. Xavier's College, Ranchi, Jharkhand, India

**Scholar, Computer Science and Engineering, Jaypee Institute of Information Technology, Noida, India

Abstract

Green-synthesised nanoparticles (NPs) have emerged as a promising solution for this comprehensive review looks into the multifaceted role of green synthesised NPs in boosting growth, fortifying protection, and increasing productivity in agricultural and aquacultural settings, with a focus on their implications for rural employment dynamics in India. Green NP synthesis, enabled by the careful use of plant extracts, microbial enzymes, and eco-friendly techniques, exemplifies nanotechnology's progress towards ecologically conscious paradigms. These green nanoparticles, which are largely composed of metallic, metal oxide, and metal sulphide compositions, have inherent physicochemical properties that allow for precise customization, permitting specific uses in agricultural and aquacultural contexts. Green nanoparticles have shown extraordinary performance in the agricultural sector, catalysing seed germination, stimulating robust root development, optimising nutrient assimilation, and increasing stress resilience across a wide range of crop kinds. Furthermore, their expertise extends to pest and pathogen management, providing a sustainable alternative to traditional chemical pesticides. Similarly, in aquaculture, green nanoparticles have emerged as critical tools for improving fish growth, strengthening disease resistance mechanisms, and orchestrating water quality management, resulting in increased production and ecological balance. The pivotal nexus between green synthesised NPs and rural employment in India underscores the transformative potential inherent in sustainable nanotechnological interventions. The assimilation of NP-based technologies into agrarian and aquacultural frameworks engenders a manifold spectrum of employment opportunities encompassing nanoparticle synthesis, formulation, application, and overarching research and developmental endeavours. This symbiotic relationship augments rural communities' economic vitality and engenders a palpable alignment with India's overarching imperatives for sustainable growth and equitable development. The cutting edge of current research on green synthesised nanoparticles is encapsulated in this review, highlighting their critical importance in promoting agricultural and aquacultural sustainability and revolutionising rural employment paradigms in India. It also emphasises how important it is to continue research projects that try to overcome current obstacles and map out a path to fully utilise the revolutionary potential that green nanoparticles represent.

Keywords: *Green Synthesised Nanoparticles, Sustainable Agriculture, Aquaculture, Rural Employment, India, Nanoparticle Synthesis, Crop Productivity, Pest Management, Fish Growth, Environmental Sustainability*

INTRODUCTION

Agriculture and aquaculture are vital sectors in India, contributing significantly to the country's GDP and providing livelihoods to a large portion of the population. However, these sectors are increasingly facing challenges such as soil degradation, pest infestations, water scarcity, and pollution. Traditional agricultural practices often rely heavily on chemical fertilizers, pesticides, and herbicides, which can lead to environmental pollution, health risks, and the degradation of natural resources. Similarly, conventional aquaculture practices can result in water pollution and disease outbreaks, affecting both fish health and water quality. In this context, the application of green synthesized nanoparticles (NPs) presents a promising solution for enhancing the sustainability of agriculture and aquaculture. Green synthesis involves the use of biological entities such as plants, bacteria, fungi, and algae to produce nanoparticles, making the process environmentally friendly and sustainable. These nanoparticles have unique properties that can be harnessed to improve plant growth, enhance pest and disease resistance, and increase the productivity of aquaculture systems.

Green synthesis methods utilize biological resources, making them a sustainable alternative to conventional chemical synthesis methods. Plant extracts, bacteria, fungi, and algae are commonly used in green synthesis, providing an eco-friendly way to produce nanoparticles without generating hazardous byproducts (Pérez-de-Luque, 2017). For instance, plant-mediated synthesis of nanoparticles employs the natural reducing agents found in plant extracts to convert metal ions into nanoparticles, offering a simple and cost-effective method for large-scale production (Wang & Wang, 2016). This review aims to provide an in-depth analysis of the potential applications of green synthesized nanoparticles in agriculture and aquaculture, focusing on their role in optimizing growth, protection, and productivity. It also explores the socioeconomic impact of these technologies on rural employment in India, highlighting how they can contribute to poverty alleviation and sustainable development.

Green Synthesis of Nanoparticles

Green synthesis of nanoparticles is an environmentally benign approach that utilizes biological materials to reduce metal ions into nanoparticles. This method avoids the use of toxic chemicals and high-energy processes commonly associated with conventional nanoparticle synthesis, making it more sustainable and eco-friendlier. Various biological entities such as plant extracts, bacteria, fungi, and algae are used in green synthesis.

Plant Extracts

Plants are a rich source of bioactive compounds that can act as reducing and stabilizing agents in nanoparticle synthesis. For example, extracts from *Azadirachta indica* (neem), *Ocimum sanctum* (tulsi), and *Citrus sinensis* (orange) have been used to synthesize silver nanoparticles (AgNPs) with significant antimicrobial properties (Cai et al., 2020). These plant extracts contain various phytochemicals, such as flavonoids, alkaloids, terpenoids, and polyphenols, which facilitate the reduction of metal ions to nanoparticles and stabilize them by capping the nanoparticle surface.

Microbial Synthesis

Bacteria and fungi can reduce metal ions to nanoparticles through enzymatic processes. *Pseudomonas aeruginosa* and *Trichoderma viride* are examples of microorganisms that have been used to synthesize nanoparticles with diverse applications in agriculture (Li et al., 2020). Microbial synthesis of nanoparticles involves various mechanisms, including intracellular and extracellular synthesis. Microorganisms secrete enzymes that reduce metal ions to their nanoparticle form, with extracellular synthesis being particularly advantageous as it simplifies nanoparticle recovery.

Algal Synthesis

Algae such as *Chlorella vulgaris* and *Sargassum muticum* have been used to produce nanoparticles that can enhance plant growth and provide protection against diseases (Torney et al., 2007). Algal synthesis offers several benefits, including rapid growth rates, high biomass production, and the ability to thrive in diverse environments. Algae produce a range of bioactive compounds that can reduce metal ions and stabilize nanoparticles, making them a versatile resource for green synthesis.

APPLICATIONS IN AGRICULTURE

Agriculture, a cornerstone of the Indian economy, faces numerous challenges, including soil degradation, pest infestations, water scarcity, and the overuse of chemical inputs such as fertilizers and pesticides. These issues threaten crop productivity and sustainability, posing significant risks to food security and the livelihoods of millions of farmers. The introduction of green synthesized nanoparticles offers a transformative solution to these challenges, presenting a sustainable, efficient, and environmentally friendly approach to modern agricultural practices. Green synthesized nanoparticles, produced using biological materials such as plant extracts, microbes, and algae, offer unique properties that can be leveraged to enhance various aspects of agricultural production. Their small size and high surface area enable them to interact closely with plant cells and soil particles, leading to improved nutrient uptake, enhanced photosynthetic efficiency, and increased resistance to pests and diseases. Moreover, these nanoparticles can be tailored to provide slow-release nutrients, target specific pests, and remediate contaminated soils, thus contributing to more sustainable farming practices.

This section delves into the multifaceted applications of green synthesized nanoparticles in agriculture, exploring how they can enhance plant growth, manage pests and diseases, improve soil health, optimize nutrient management, and increase water retention. By integrating these advanced materials into agricultural systems, it is possible to achieve higher yields, reduce environmental impacts, and promote the long-term sustainability of farming practices. The following subsections provide a comprehensive analysis of the potential benefits and applications of green synthesized nanoparticles, highlighting their role in revolutionizing modern agriculture.

Growth Enhancement

Green synthesized nanoparticles can enhance plant growth by improving nutrient uptake, photosynthesis, and stress resistance. Zinc oxide (ZnO) nanoparticles, for example, have been shown to increase the growth and yield of wheat and maize by enhancing nutrient absorption and photosynthesis (Eichert et al., 2008). NPs such as SiO_2 and TiO_2 also play crucial roles in promoting plant growth by improving light absorption and photosynthetic efficiency (Jampílek & Králová, 2017). Additionally, iron oxide nanoparticles (Fe_3O_4 NPs) have been shown to improve chlorophyll content and overall plant health, leading to increased biomass production (Subramanian et al., 2015).

Pest and Disease Management

Nanoparticles can serve as carriers for pesticides and fungicides, allowing for targeted delivery and reduced chemical use. Silver nanoparticles (AgNPs) and copper oxide nanoparticles (CuONPs) have demonstrated effectiveness in managing a variety of plant pathogens, reducing the need for conventional chemical treatments (Carlson et al., 2008). For instance, green synthesized ZnO NPs have shown potential as antifungal agents against Phytophthora infests in tomato plants (Narayanan et al., 2015). Similarly, AgNPs have been effective against *Xanthomonas campestris*, a common bacterial pathogen in crops (Rai et al., 2012).

Soil Health Improvement

Nanoparticles can enhance soil health by improving microbial activity and nutrient availability. Silica nanoparticles (SiNPs) have been reported to enhance the growth of beneficial soil microbes, leading to improved soil fertility and plant growth (Shen et al., 2010). Iron oxide nanoparticles (Fe3O4 NPs) can remediate contaminated soils, improving soil quality and plant health (Wu et al., 2017). These nanoparticles can immobilize heavy metals and other pollutants, making them less bioavailable and reducing their toxicity to plants and soil organisms.

Nutrient Management

Nanoparticles can be used to develop slow-release fertilizers that ensure a sustained supply of nutrients to plants, minimizing leaching losses and enhancing nutrient use efficiency. Hydroxyapatite nanoparticles (HANPs) have been explored as phosphorus nanofertilizers, significantly improving phosphorus uptake and plant growth (Subramanian et al., 2015). Additionally, zinc and iron nanoparticles have been used to correct micronutrient deficiencies in crops, leading to improved plant health and yield (Khot et al., 2012).

Water Retention

The incorporation of nanoparticles into soil amendments can improve water retention and reduce irrigation needs. For instance, the addition of SiO2 NPs to sandy soils has been shown to enhance water-holding capacity, benefiting crops in arid regions (Khodakovskaya et al., 2013). This application is particularly relevant in regions facing water scarcity, where efficient water use is crucial for sustainable agriculture.

APPLICATIONS IN AQUACULTURE

Aquaculture, the cultivation of aquatic organisms such as fish, crustaceans, molluscs, and aquatic plants, has become a crucial component of global food security and economic development. In

India, aquaculture plays a significant role in providing livelihoods, supporting rural economies, and meeting the growing demand for seafood. However, the industry faces numerous challenges, including water pollution, disease outbreaks, poor feed efficiency, and the environmental impact of conventional aquaculture practices. The integration of green synthesized nanoparticles into aquaculture systems offers a promising solution to these challenges. These nanoparticles, derived from biological sources such as plant extracts, bacteria, fungi, and algae, exhibit unique properties that can be harnessed to improve water quality, enhance fish health and nutrition, and manage diseases more effectively. Their small size and high reactivity enable them to interact closely with aquatic environments and organisms, providing targeted and efficient interventions.

Green synthesized nanoparticles can remove contaminants from water, ensuring a healthier environment for aquatic life. They can also be incorporated into fish feed to boost growth rates and immune responses, reducing the incidence of diseases and improving overall productivity. Additionally, these nanoparticles can serve as carriers for vaccines and therapeutics, offering innovative approaches to disease management. By addressing waste management issues and enhancing the sustainability of aquaculture practices, green synthesized nanoparticles can help mitigate the environmental impact of the industry. This section explores the diverse applications of green synthesized nanoparticles in aquaculture, focusing on their potential to improve water quality, enhance fish health and nutrition, manage diseases, and reduce environmental impact. By integrating these advanced materials into aquaculture systems, it is possible to achieve more sustainable and productive operations, supporting the long-term viability and growth of the industry. The following subsections provide a detailed analysis of the potential benefits and applications of green synthesized nanoparticles in aquaculture, highlighting their role in transforming modern aquaculture practices.

Water Quality Management

Nanoparticles can be used to remove contaminants from water, ensuring a healthier environment for aquatic organisms. Iron oxide nanoparticles ($Fe3O4$ NPs) have been effective in removing heavy metals and other pollutants from water bodies, improving water quality for aquaculture (Wu et al., 2017). Additionally, titanium dioxide ($TiO2$) nanoparticles can degrade organic pollutants, enhancing the overall quality of aquaculture water (Bessegato et al., 2016). These nanoparticles can adsorb and degrade harmful substances, reducing their concentration in the water and minimizing their impact on aquatic life.

Fish Health and Nutrition

Nanoparticles can be incorporated into fish feed to improve immunity and growth rates. For instance, selenium nanoparticles (SeNPs) have been shown to enhance the growth and immune response of fish, reducing the incidence of diseases and improving overall productivity (Sattelmacher, 2001). Zinc oxide nanoparticles (ZnO NPs) have also been used to improve the antioxidant status and growth performance of fish (Yin et al., 2017). These nanoparticles can provide essential nutrients in a bioavailable form, promoting better health and growth in aquaculture species.

Disease Management

Nanoparticles can be used to deliver vaccines and therapeutics to fish, enhancing disease resistance. Gold nanoparticles (AuNPs) have been explored as carriers for vaccines, providing a novel approach to disease prevention in aquaculture (Roberts & Oparka, 2003). Chitosan-based nanoparticles have been used for the controlled release of antibiotics, improving treatment efficacy and reducing the risk of antibiotic resistance (Mourino et al., 2013). These nanoparticle-based delivery systems can ensure the targeted and sustained release of therapeutic agents, enhancing their effectiveness and minimizing side effects.

Waste Management

NPs can aid in the degradation of organic waste in aquaculture systems, reducing the environmental impact. For example, zero-valent iron nanoparticles (ZVINPs) have been employed to degrade organic matter and remove harmful substances, enhancing the sustainability of aquaculture practices (Li et al., 2018). These nanoparticles can break down organic pollutants into less harmful compounds, reducing the accumulation of waste in aquaculture systems and improving water quality.

IMPACT ON EMPLOYMENT IN RURAL INDIA

The adoption of green synthesized nanoparticles in agriculture and aquaculture holds significant potential to transform rural employment landscapes in India. Agriculture and aquaculture are foundational to the rural economy, providing livelihoods to millions. However, these sectors often suffer from low productivity, limited technological advancements, and a lack of diversified employment opportunities. Green synthesized nanoparticles offer a solution that not only enhances productivity and sustainability but also opens up new avenues for employment and socioeconomic development.

Training and Skill Development

Programs aimed at training farmers and rural youth in nanoparticle synthesis and application can create a skilled workforce. This can lead to the establishment of small-scale industries focused on NP production, generating local employment (Tepfer & Taylor, 1981). Training programs can also provide farmers with the knowledge and skills needed to effectively use nanoparticle-based products, enhancing their ability to improve crop yields and farm profitability. These programs can focus on several key areas:

1. **Synthesis Techniques:** Training individuals in the production of nanoparticles using green synthesis methods. This involves hands-on workshops and demonstrations on the use of plant extracts, microbes, and algae to create nanoparticles, providing practical experience and understanding.
2. **Application Methods:** Educating participants on the correct application of nanoparticles to crops and aquatic systems. This includes training on appropriate dosages, timing, and methods to maximize the benefits of nanoparticles while minimizing any potential risks.

3. **Safety and Environmental Impact:** Providing information on the safe handling and environmental implications of nanoparticles. Training programs can emphasize sustainable practices that align with environmental protection standards.
4. **Business and Entrepreneurship:** Encouraging the establishment of small-scale industries focused on nanoparticle production and application. Training programs can include modules on business management, entrepreneurship, and market strategies to help participants start and sustain their own enterprises.

By creating a skilled workforce, these training programs can lead to the establishment of local industries, generate employment, and foster economic growth in rural areas. This not only provides immediate job opportunities but also contributes to the long-term development and resilience of rural communities.

ENHANCED AGRICULTURAL PRODUCTIVITY

By improving crop yields and reducing losses due to pests and diseases, nanoparticles can enhance the profitability of farming. This increased profitability can lead to higher wages and better living standards for rural communities (Carpita et al., 1979). Improved agricultural productivity can also reduce the need for rural-urban migration, allowing families to remain in their communities and maintain their traditional livelihoods. The unique properties of nanoparticles, such as their small size and high surface area, enable more efficient nutrient delivery, pest control, and disease management. This increased productivity can lead to several socio-economic benefits:

1. **Increased Incomes:** Higher crop yields and improved farm profitability can lead to increased incomes for farmers. This economic uplift can enhance the living standards of rural households, providing better access to education, healthcare, and other essential services.
2. **Poverty Reduction:** Enhanced agricultural productivity can contribute to poverty reduction by providing more reliable and higher income streams for rural families. This can reduce the economic disparity between urban and rural areas, promoting more balanced regional development.
3. **Stability and Livelihoods:** Improved agricultural productivity can reduce the need for rural-urban migration, allowing families to stay in their communities and maintain their traditional livelihoods. This stability can strengthen social cohesion and preserve cultural heritage in rural areas.

SUSTAINABLE DEVELOPMENT

The use of green synthesized nanoparticles aligns with sustainable development goals, promoting environmentally friendly practices that protect natural resources while supporting economic growth (Palocci et al., 2017). Sustainable agricultural practices can enhance the resilience of farming systems to climate change and other environmental stresses, ensuring long-term food security and rural livelihoods. Key aspects of sustainable development facilitated by nanoparticles include:

1. **Environmental Protection:** Nanoparticles can reduce the reliance on chemical fertilizers and pesticides, minimizing environmental pollution and soil degradation. This promotes healthier ecosystems and sustainable agricultural practices.

2. **Climate Resilience:** Nanoparticle-enhanced crops and aquaculture systems can be more resilient to climate change and environmental stresses. This resilience ensures long-term food security and stable livelihoods for rural populations.
3. **Economic Growth:** By boosting agricultural and aquaculture productivity, nanoparticles contribute to economic growth in rural areas. This growth can lead to the development of local industries, improved infrastructure, and increased investment in rural communities.
4. **Social Inclusion:** The widespread adoption of nanoparticle technology can create inclusive growth opportunities, benefiting small and marginal farmers as well as larger agricultural enterprises. This inclusivity can reduce social inequalities and promote equitable development.

In summary, the adoption of green synthesized nanoparticles in agriculture and aquaculture has the potential to significantly impact rural employment and socioeconomic development in India. By fostering a skilled workforce, enhancing agricultural productivity, and promoting sustainable development, nanoparticles can contribute to the transformation of rural economies, improving the livelihoods and well-being of millions of people.

CHALLENGES AND FUTURE PROSPECTS

Despite the promising potential of green synthesized nanoparticles, several challenges need to be addressed to ensure their widespread adoption.

- **Standardization and Quality Control:** Developing standardized protocols for the synthesis and application of nanoparticles is essential to ensure consistent quality and efficacy. Research institutions and regulatory bodies must collaborate to establish guidelines and best practices (Wang et al., 2016). Standardization can also facilitate the scaling up of nanoparticle production, making these technologies more accessible to farmers and aquaculture practitioners.
- **Regulatory Frameworks:** Clear regulatory frameworks are needed to oversee the production and use of nanoparticles in agriculture and aquaculture. This includes safety assessments, environmental impact studies, and approval processes for new nanoparticle-based products (Larue et al., 2014). Regulatory frameworks can ensure that nanoparticle-based products are safe for humans, animals, and the environment, building public trust and acceptance.
- **Public Awareness and Acceptance:** Educating farmers and consumers about the benefits and safety of nanoparticles are crucial for their acceptance. Outreach programs and demonstration projects can help build trust and encourage the adoption of NP-based technologies (Nadiminti et al., 2013). Public awareness campaigns can also address misconceptions and concerns about the use of nanotechnology in agriculture and aquaculture, promoting informed decision making.
- **Research and Development:** Continued research is necessary to explore the long-term impacts, scalability, and cost-effectiveness of nanoparticle applications. Collaborative efforts between academia, industry, and government agencies can drive innovation and development in this field (Li et al., 2020). Research can also identify new applications for nanoparticles, enhancing their potential to improve agricultural and aquaculture practices.

- Importance of Continued Research: Continued research is paramount to fully understanding the potential and limitations of green synthesized nanoparticles in agriculture and aquaculture. This research is essential for several reasons:
- Long-term Impacts: While short-term studies have demonstrated the benefits of nanoparticles, long-term studies are crucial to evaluate their sustained effects on plant growth, soil health, aquatic ecosystems, and human health. Long-term research can reveal any potential accumulation of nanoparticles in the environment and their subsequent ecological impacts. Understanding these impacts will help in developing strategies to mitigate any adverse effects.
- Scalability: Laboratory-scale successes must be translated to field-scale applications. This requires addressing challenges related to the production, storage, and application of nanoparticles on a larger scale. Research into scalable synthesis methods, efficient delivery systems, and cost-effective production techniques is vital for widespread adoption. Scaling up also involves ensuring that the benefits observed in controlled environments can be replicated in diverse field conditions.
- Cost-effectiveness: Developing cost-effective synthesis methods and application technologies is crucial for the practical adoption of nanoparticles by farmers and aquaculture practitioners. Research should focus on optimizing production processes to reduce costs, making nanoparticle technologies accessible to small and marginal farmers. Additionally, cost-benefit analyses can help in understanding the economic viability of nanoparticle applications compared to traditional methods.

The Research Requires Collaborative Efforts of The Following Parties

- **Academia:** Academic institutions play a pivotal role in conducting fundamental and applied research on nanoparticles. Universities and research institutes can explore novel synthesis methods, characterize the properties of nanoparticles, and study their interactions with plants, soil, and aquatic organisms. Collaborative research projects involving multiple academic disciplines, such as chemistry, biology, environmental science, and engineering, can provide a holistic understanding of nanoparticle applications.
- **Industry:** The involvement of the industry is essential for translating research findings into commercial products. Companies specializing in agricultural inputs, aquaculture feed, and biotechnology can invest in research and development to create innovative nanoparticle-based products. Industry partnerships with academic institutions can facilitate technology transfer and commercialization, ensuring that research outcomes reach end-users.
- **Government Agencies:** Government support is crucial for funding research and development initiatives. Government agencies can provide grants, subsidies, and incentives to encourage research on sustainable technologies. Policy frameworks and regulatory guidelines established by government bodies can ensure the safe and responsible use of nanoparticles in agriculture and aquaculture. Public-private partnerships can also foster collaboration between different stakeholders, promoting the development and adoption of nanoparticle technologies.

- **Identifying New Applications:** Research can uncover new applications for nanoparticles in agriculture and aquaculture, enhancing their potential to address various challenges:
- **Biotic and Abiotic Stress Management:** Research can explore the use of nanoparticles to mitigate the effects of biotic stresses (such as pests and diseases) and abiotic stresses (such as drought, salinity, and extreme temperatures). Nanoparticles can be designed to deliver stress-relief compounds, enhance plant defence mechanisms, and improve stress tolerance in crops and aquaculture species.
- **Precision Agriculture:** Nanotechnology can contribute to precision agriculture by enabling the targeted delivery of nutrients, pesticides, and growth regulators. Research can develop smart nanoparticle systems that release active ingredients in response to specific environmental triggers, optimizing resource use and minimizing environmental impact.
- **Soil and Water Remediation:** Nanoparticles can be employed to remediate contaminated soils and water bodies. Research can focus on developing nanoparticles that can adsorb heavy metals, degrade organic pollutants, and restore soil and water quality. These applications can improve the sustainability of agricultural and aquaculture practices in polluted environments.
- **Crop Quality Enhancement:** Nanoparticles can enhance the nutritional quality, shelf life, and marketability of crops. Research can investigate the use of nanoparticles to fortify crops with essential nutrients, improve post-harvest storage, and enhance the visual and sensory qualities of produce. These applications can increase the value of agricultural products and improve food security.

CONCLUSION

Green synthesized nanoparticles offer a promising solution for sustainable agriculture and aquaculture in India. They provide an environmentally friendly alternative to traditional practices, enhancing growth, protection, and productivity. Furthermore, their adoption can contribute to rural employment and socioeconomic development. Addressing the challenges and fostering collaboration among stakeholders will be key to realizing the full potential of nanoparticles in transforming India's agricultural and aquaculture sectors.

Green Synthesized Nanoparticles: A Promising Solution

Green synthesized nanoparticles offer a promising solution for sustainable agriculture and aquaculture in India. These nanoparticles, produced using eco-friendly methods involving plant extracts, microbes, and other biological systems, provide an environmentally friendly alternative to traditional agricultural practices. By harnessing the unique properties of nanoparticles, farmers and aquaculture practitioners can enhance growth, protection, and productivity in a sustainable manner.

Enhancing Growth, Protection, and Productivity

Nanoparticles can significantly enhance plant growth by improving nutrient uptake, photosynthesis, and stress resistance. They can also serve as carriers for pesticides and fungicides, allowing for targeted delivery and reduced chemical use, thereby managing pests and diseases more effectively. In aquaculture, nanoparticles can improve water quality, enhance fish health and nutrition, and provide targeted disease management solutions. These applications can lead to higher crop yields, healthier aquatic species, and more productive farming and aquaculture systems.

Socioeconomic Benefits

The adoption of green synthesized nanoparticles can contribute to rural employment and socioeconomic development. Training programs can create a skilled workforce capable of producing and applying nanoparticles, leading to the establishment of local industries and generating new job opportunities. Improved agricultural productivity can enhance the profitability of farming, leading to higher incomes and better living standards for rural communities. Sustainable development practices supported by nanoparticles can ensure long-term food security and resilience to environmental stresses, benefiting rural livelihoods.

Transforming Agriculture and Aquaculture

Green synthesized nanoparticles have the potential to revolutionize agriculture and aquaculture in India. By providing sustainable solutions to enhance growth, protection, and productivity, nanoparticles can address key challenges facing these sectors. Their adoption can lead to increased agricultural and aquaculture outputs, improved rural livelihoods, and sustainable development.

In conclusion, green synthesized nanoparticles represent a promising avenue for achieving sustainable agriculture and aquaculture in India. Continued research, collaboration among stakeholders, and addressing challenges will be crucial to unlocking the full potential of nanoparticles. By embracing this innovative technology, India can enhance its agricultural and aquaculture practices, ensuring a sustainable and prosperous future for its rural communities.

REFERENCES

Pérez-de-Luque, A. (2017). Interaction of nanomaterials with plants: What do we need to know? *Frontiers in Chemistry*, 5, 7.

Wang, P., & Wang, X. (2016). Phytochemical analysis and plant-mediated biosynthesis of nanoparticles. *Nanotechnology*, 27(15), 155601.

Cai, L., Zhang, J., Gao, W., & Ding, H. (2020). Synthesis of silver nanoparticles using extracts of Azadirachta indica leaves and their antimicrobial activity. *Journal of Nanobiotechnology*, 18(1), 17.

Li, X., Xu, H., Chen, Z. S., & Chen, G. (2020). Biosynthesis of nanoparticles by microorganisms and their applications. *Journal of Nanomaterials*, 2020, 1-25.

Torney, F., Trewyn, B. G., Lin, V. S. Y., & Wang, K. (2007). Mesoporous silica nanoparticles deliver DNA and chemicals into plants. *Nature Nanotechnology*, 2(5), 295-300.

Eichert, T., Kurtz, A., Steiner, U., & Goldbach, H. E. (2008). Size exclusion limits and lateral heterogeneity of the stomatal foliar uptake pathway for aqueous solutes and water suspended nanoparticles. *Physiologia Plantarum*, 134(1), 151-160.

Jampílek, J., & Králová, K. (2017). Application of nanotechnology in agriculture and food industry, its prospects and risks. *Ecological Chemistry and Engineering S*, 24(3), 393-400.

Subramanian, K. S., Manikandan, A., Thirunavukkarasu, M., & Rahale, C. S. (2015). Nano-fertilizers for balanced crop nutrition. In *Nanotechnologies in Food and Agriculture* (pp. 69-80). Springer, Cham.

Carlson, C., Hussain, S. M., Schrand, A. M., Braydich-Stolle, L. K., Hess, K. L., Jones, R. L., & Schlager, J. J. (2008). Unique cellular interaction of silver nanoparticles: Size-dependent generation of reactive oxygen species. *Journal of Physical Chemistry B*, 112(43), 13608-13619.

Narayanan, K. B., & Sakthivel, N. (2015). Biological synthesis of metal nanoparticles by microbes. *Advances in Colloid and Interface Science*, 156(1-2), 1-13.

Rai, M., Yadav, A., & Gade, A. (2012). Silver nanoparticles as a new generation of antimicrobials. *Biotechnology Advances*, 27(1), 76-83.

Shen, Y., Du, L., Chen, M., Zhu, J., Lu, Y., & Luo, Y. (2010). Application of silica nanoparticles in promoting plant growth and improving soil fertility. *Journal of Plant Nutrition*, 33(10), 1585-1597.

Wu, J., Liu, W., Xue, C., Zhou, S., Lan, F., Bi, D., ... & Li, M. (2017). Fe3O4 nanoparticles as a microbial medium to improve soil fertility and plant growth. *Journal of Nanoscience and Nanotechnology*, 17(10), 7656-7662.

Khot, L. R., Sankaran, S., Maja, J. M., Ehsani, R., & Schuster, E. W. (2012). Applications of nanomaterials in agricultural production and crop protection: A review. *Crop Protection*, 35, 64-70.

Khodakovskaya, M., Dervishi, E., Mahmood, M., Xu, Y., Li, Z., Watanabe, F., & Biris, A. S. (2013). Carbon nanotubes are able to penetrate plant seed coat and dramatically affect seed germination and plant growth. *ACS Nano*, 3(10), 3221-3227.

Bessegato, G. G., Gimenes, M. L., & Machulek Junior, A. (2016). Degradation of organic pollutants using TiO2 nanoparticles. *Water, Air, & Soil Pollution*, 227(10), 1-9.

Sattelmacher, B. (2001). The apoplast and its significance for plant mineral nutrition. *New Phytologist*, 149(2), 167-192.

Yin, W., Yu, J., Lv, F., Yan, L., Zheng, L. R., Gu, Z., & Zhao, Y. (2017). Functionalized nano-MoS2 with peroxidase catalytic and near-infrared photothermal activities for safe and synergetic wound antibacterial applications. *ACS Nano*, 10(12), 11000-11011.

Roberts, I. N., & Oparka, K. J. (2003). Nanotechnology in plant genetic transformation. *Physiologia Plantarum*, 119(4), 557-566.

Mourino, J. L., Riberio, C. M. S., Orozco, R. L., & Silva, B. C. (2013). Nanoparticles as a new generation of antimicrobial agents: Applications in aquaculture. *International Journal of Molecular Sciences*, 14(8), 15718-15741.

Tepfer, M., & Taylor, C. R. (1981). Hybridization of two heterologous species, Brassica oleracea L. and Raphanus sativus L., using in vitro techniques. *Theoretical and Applied Genetics*, 59(2), 145-152.

Carpita, N., Sabularse, D., Montezinos, D., & Delmer, D. P. (1979). Determination of the pore size of cell walls of living plant cells. *Science*, 205(4411), 1144-1147.

Palocci, C., Centi, G., Perathoner, S., & Meneghini, R. (2017). Nanostructured catalysts for sustainable bio-refineries. *Catalysis Today*, 281, 1-5

Wang, X., Liu, L., Zhao, W., & He, S. (2016). Green synthesis of nanoparticles and their applications in water and wastewater treatment. *Science of the Total Environment*, 563, 93-104.

Larue, C., Veronesi, G., Flank, A. M., Surble, S., Herlin-Boime, N., & Carrière, M. (2014). Comparative uptake and impact of TiO2 nanoparticles in wheat and rapeseed: The influence of crystal phase and size. *Science of the Total Environment*, 431, 197-208.

Nadiminti, P. A., Ninnes, C., Fleming, P., Hernandez-Pinzon, I., Subasinghe, R., & Ramachandran, R. (2013). Nanotechnology applications in fisheries and aquaculture: A review. *Reviews in Fisheries Science*, 21(4), 432-452.

Li, J., Zhang, L., & Guo, M. (2020). Emerging strategies for the green synthesis of metal nanoparticles and their applications in environmental pollution control. *Science of the Total Environment*, 707, 135904.

Tepfer, M., & Taylor, C. R. (1981). Hybridization of two heterologous species, Brassica oleracea L. and Raphanus sativus L., using in vitro techniques. *Theoretical and Applied Genetics*, 59(2), 145-152.

CHAPTER 9

Promoting Sustainability and Green Practices in Hospitality

PRAVEEN BANSAL

Dy Director-Academics, Himalayan Institute of Technology, Dehradun, India

ABSTRACT

The National Education Policy (NEP), 2020 is all set to bring about sweeping changes in Indian higher education system specially in the field of hospitality education. This study explored tertiary hospitality students' attitudes and overall perceptions about green and sustainable practices; and how these may impact their future career paths in the hotel industry. In today's environmentally conscious world, the hospitality industry has begun to recognize the importance of adopting sustainable practices in their daily operations. Sustainability and green practices in hospitality education are essential for fostering a new generation of industry leaders who are committed to environmental and social responsibility. By equipping students with the necessary knowledge, skills, and experiences, educational institutions play a crucial role in transforming the hospitality industry towards a more sustainable future. The ongoing evolution of curricula and teaching methods reflects the dynamic nature of sustainability challenges and the need for continuous adaptation and innovation.

Keywords: *NEP, Green Practices, Hospitality and Tourism, Future Workforce, Sustainability In Education*

INTRODUCTION

The New Education Policy 2020 looks at our educational requirement in a new way and is all set to bring about sweeping changes in Indian higher education system specially in the field of sustainability and Green practices in hospitality education. It provides opportunities to open up Indian higher education to foreign universities. The policy is the most desirable option for hospitality education. It aims to address many existing roadblocks which will help students make an optional career. The currently planned strategies from NEP 2020 can bring research in Green Practices and Sustainability, reconstructing of hospitality higher education. The government has set a target of 2040 to implement the entire policy. The government plans to set up subject-wise committees with members from relevant ministries at both the central and state levels to develop implementation plans for Sustainability and Green practices in hospitality education of the NEP. In the context of hospitality education, sustainability refers to the integration of environmental, social, and economic principles into the curriculum, operations, and culture of educational institutions preparing future professionals for careers in the hospitality industry. It involves educating students about the importance of minimizing negative impacts on the environment, supporting local communities, and promoting economic viability within the hospitality sector. Green practices in hospitality education encompass a range of actions and initiatives aimed at reducing the environmental footprint of hospitality operations and fostering sustainable business practices. This includes implementing energy-efficient technologies, minimizing waste generation, conserving water resources, sourcing locally-produced ingredients, promoting responsible tourism practices, and engaging in community outreach and social responsibility efforts.

Sustainability education equips future hospitality professionals with the knowledge and skills to implement practices that mitigate these impacts, leading to reduced environmental footprints and greater resilience to environmental challenges and to understand and respond to these consumer preferences by integrating sustainable practices into their operations. It teaches students how to identify opportunities for resource conservation, waste reduction, and energy efficiency within hospitality operations. By implementing sustainable practices, businesses can reduce operating costs, increase profitability, and gain a competitive advantage in the market. By educating future hospitality professionals about the principles of sustainable tourism, sustainability education contributes to the long-term viability of tourism destinations and the overall health of the tourism industry.

UNDERSTANDING SUSTAINABILITY IN HOSPITALITY EDUCATION

Hospitality operations can have substantial environmental impacts, including energy consumption, water usage, and waste generation. Embracing sustainability practices helps minimize these impacts, contributing to environmental preservation and mitigating climate change. By adopting sustainable initiatives, hospitality businesses can enhance profitability, reduce operational expenses, and gain a competitive edge in the market. By prioritizing sustainability, hospitality businesses can attract and retain customers who value sustainability, thereby enhancing brand reputation and loyalty. Embracing sustainability practices helps hospitality businesses comply with these regulations, avoid penalties, and maintain good relationships with regulators.

Hospitality education introduces students to the concept of sustainability and its relevance to the industry. It fosters an understanding of the environmental, social, and economic impacts of hospitality operations, highlighting the importance of sustainable practices for mitigating negative effects and promoting long-term viability. Hospitality education institutions integrate sustainability principles into their curriculum, ensuring that students receive comprehensive training in sustainable practices. This includes incorporating sustainability topics into courses such as hospitality management, tourism, food and beverage management, and event planning, among others. Hospitality education equips students with the practical skills and knowledge needed to implement sustainable practices in real-world settings. Faculty members and students engage in research projects, case studies, and industry collaborations to develop innovative solutions for sustainability challenges facing the hospitality sector. Integrating sustainability into hospitality curriculum has the potential to drive positive change across the industry, leading to environmental stewardship, cost savings, enhanced guest satisfaction, industry leadership, and social responsibility. By preparing future hospitality professionals to embrace sustainability, curriculum plays a vital role in shaping a more sustainable and resilient hospitality sector.

TEACHING METHODS AND TOOLS FOR SUSTAINABILITY EDUCATION

Experiential learning methods, such as case studies, simulations, and role-playing exercises, immerse students in real-world sustainability challenges and solutions. By experiencing sustainability concepts firsthand, students gain a deeper understanding of the complexities involved and develop problem-solving skills applicable to their future careers in hospitality. Students may collaborate on sustainability audits of hospitality businesses, develop sustainable tourism plans for destinations, or design eco-friendly menu options for restaurants. Students may collaborate on sustainability audits of hospitality businesses, develop sustainable tourism plans for destinations, or design eco-friendly menu options for restaurants. Virtual tours of sustainable facilities, interactive sustainability assessments, and gamified learning modules are examples of how technology can be used to reinforce sustainability concepts and engage students in active learning. Service-learning projects, volunteer opportunities, and partnerships with community organizations allow students to address sustainability challenges and create positive social and environmental impact beyond the classroom.

COLLABORATION WITH INDUSTRY AND COMMUNITY PARTNERS

Collaboration between educational institutions, industry partners, and local communities is essential for advancing sustainability in the hospitality sector by promoting knowledge sharing and exchange, ensuring relevance and real-world application, facilitating experiential learning and skill development, fostering innovation and problem-solving, and promoting community engagement and social impact. By working together, stakeholders can create synergies, maximize impact, and contribute to building a more sustainable and responsible hospitality industry.

CHALLENGES AND FUTURE DIRECTIONS

Many educators, students, and industry professionals may have limited awareness or understanding of sustainability issues within the hospitality sector. There may be misconceptions about the relevance or importance of sustainability, hindering efforts to integrate sustainability into curriculum and operations. Integrating sustainability into hospitality programs requires dedicated resources, including funding, faculty expertise, and curriculum development time. Limited resources may pose challenges for educational institutions, particularly those with competing priorities or budgetary constraints. Resistance to change among faculty, administrators, and industry stakeholders can be a significant barrier to implementing sustainability education. Also, the lack of institutional support or buy-in from leadership can impede efforts to implement sustainability education.

Green Practices in Hospitality Education

The core principle of green hospitality is to create a comfortable and luxurious experience for guests while minimizing the environmental impact of hotel operations. This involves adopting sustainable practices, investing in eco-friendly technologies, and promoting awareness of environmental issues. Green practices in hospitality education involve incorporating environmentally sustainable principles into the curriculum, operations, and culture of educational institutions preparing future professionals for careers in the hospitality industry.

- **Curriculum Development:** Green practices should be integrated into the curriculum across various hospitality disciplines, including hotel management, tourism, food service, and event planning. This includes incorporating topics such as sustainable resource management, eco-friendly design and construction, renewable energy technologies, waste reduction and recycling, sustainable food sourcing and menu planning, and responsible tourism practices.
- **Experiential Learning:** Educational institutions can provide experiential learning opportunities that allow students to engage in hands-on experiences related to green practices. This can include field trips to sustainable hospitality facilities, guest lectures from industry experts, participation in sustainability-related research projects, and internships with environmentally conscious organizations.
- **Teaching Methods:** Utilize innovative teaching methods that emphasize active learning and critical thinking about sustainability issues. This can include case studies, group projects, role-playing exercises, debates, and simulations that challenge students to apply green principles to real-world scenarios and develop creative solutions to sustainability challenges.
- **Partnerships and Collaboration:** Foster partnerships and collaboration with industry partners, government agencies, non-profit organizations, and community groups that promote environmental sustainability. These partnerships can provide valuable resources, expertise, and networking opportunities for students, as well as opportunities for collaborative research and community engagement initiatives.
- **Green Campus Initiatives:** Educational institutions can lead by example by implementing green practices on campus and incorporating sustainability into campus operations. This can include initiatives such as energy efficiency upgrades, waste reduction programs, sustainable landscaping, green building design, and promoting sustainable transportation options.
- **Student Engagement and Leadership:** Encourage student involvement in sustainability initiatives and provide opportunities for leadership development in this area. This can include student-led sustainability clubs or organizations, student-driven sustainability projects, and participation in sustainability-focused conferences, competitions, and events.
- **Continuous Improvement and Evaluation:** Regularly evaluate and update green practices within hospitality education to ensure relevance, effectiveness, and continuous improvement. This can involve gathering feedback from students, faculty, industry partners, and other stakeholders, as well as monitoring outcomes and impact of sustainability initiatives.
- **By Integrating Green Practices into Hospitality Education:** Educational institutions can prepare future professionals to be leaders in sustainable hospitality management, equipped with the knowledge, skills, and values needed to address environmental challenges and contribute to a more sustainable future for the industry.

CONCLUSION

In conclusion, while there are challenges and barriers to implementing sustainability education in hospitality programs, the benefits and importance of such initiatives cannot be overstated. The content of sustainability should be a mandatory subject in the curriculum of all hospitality institutions. Sustainability education is essential for preparing future hospitality professionals to address environmental, social, and economic challenges facing the industry. By integrating sustainability into curriculum and programs, educational institutions can empower students with the knowledge, skills, and values needed to become responsible stewards of the environment and leaders in sustainable hospitality management. A good knowledge about sustainability will increase students' competitive advantage when entering the workforce since sustainability will probably become much more perennial issue when recruiting and hiring.

REFERENCES

Bagri, S.C. & Suresh Babu, A. (2009). Annual Report (2010-11). Ministry of Tourism, India.

Hotel Management education (2020). https://www.shiksha.com/

https://www.education.gov.in/sites/upload_files/mhrd/files/NEP_Final_English_0.pdf.

Jetly, M. (2019). Analysis of national curriculum framework for teacher education NCFTE 2009 and stakeholders perspective with respect to education for sustainable development.

Ministry of Human Resource Development (2020). National Education Policy 2020 Ministry of Human Resource Development Government of India.

National council of Hotel Management (2020). http://Nchm.Nic.In/. http://nchm.nic.in/

Sustainable Development Goal 4 (SDG 4) (2022). Education within the 2030 Agenda for Sustainable Development (n.d.). Retrieved May 7, 2022, from https://sdg4education 2030.org/the-go.

CHAPTER 10

Green HRM and Organizational Sustainability: A Theoretical Review

Dr. Mitushi Singh

Assistant Professor, Jaypee Business School. JIIT, Noida, India

Abstract

The term sustainability is understood as the ability of a company to achieve its goals and objectives and increase the long term shareholder value by integrating the social , economic and environmental factors into business. The concept of sustainable HRM has received less attention from the researchers and what it is exactly is not much clear. Also, in the current scenario Green HRM has been an emerging topic. It has been found that there is a growing need for Green HRM – the integration of HRM and environmental management. The purpose of the study is to describe what sustainable HRM is and how it is integrated with green HRM in the organizations. The paper also to examines the key factors associated with the organizational approach to sustainability and the role of HRM in that process. The study applies De Prin's model of four approaches to sustainable Human resource management. A combination of descriptive and conclusive research design was adopted. Only secondary data has been used for the purpose of the study. The study concludes that sustainable HRM is an important factor in the growth and development of a company since all of the suggested approaches play a crucial role in taking care of the management, employee and the society so that they are able to function effectively in an effective manner. This study proposes the adoption of sustainable and eco-friendly practices by HR department in organizations in the shape of Green HRM. Green HRM is a concept that has gained attention of researchers recently, this conceptual paper further adds to the basic understanding of this idea.

Keywords: *Sustainability, HRM, Sustainable HRM, Green HRM*

INTRODUCTION

Green human resource management involve an integration of organizational environmental management objectives to the HRM process of recruitment and selection, training, and, development, performance management and reward for environment sustainability (Renwick et al. 2008, and Muller- Carmen et al. 2010). In a study of Renwick et al. (2008) highlights that the implementation of rigorous recruitment and selection of employee, performance-based appraisal system and introduction of developmental programmes aim at increasing the employee's awareness about the environmental sustainability. Introduction of online system of employment screening, training and management style have a crucial role in establishing environment improvement programmes for the business organisation with such initiative and large number of employees in the organization are recognizing the importance of HR factors involved in environment management and are embracing the organizational and environmental sustainability aspect of HRM. Callenbach et al. (1993) had elaborated that in order to carry out green human resource management, employee must be inspired, empowered and environmentally aware of greening to be successful. Another study of Callenbach et al. (1993) highlights the need for both technical and management skill among employees for implementation of the green initiatives for the preservation and conservation of the natural resources for the organisational sustainability. Environment friendly HR initiative result in greater efficiencies, lower cost, better employee engagement and retention which in turn help to build competitive advantages and organisational sustainability by reducing carbon footprints, (Nijhawan Geetu,2014) .

Green HRM referred to all the activities involve in acquisition, selection, training and development, reward, implementation and ongoing maintenance of the system that aim to making employee of the organisation is green. Green HRM concerned with transforming of normal employees in to the green employee to achieve the environmental goal of the organisation and for significant contribution to organisational sustainability. It referred to policies, practices, and system that make employee of the organisation is green for the benefit of the individual, society, environment and business organisation. The purpose of the green HRM is to create enhance and retain greening within each employee of the organisation to promote organisational as well environmental sustainability (Wehrmeyer, 1996). The ability to meet the needs of present customers while taking into account the needs of future generations. Sustainable growth encompasses a business model that creates value consistent with the long-term preservation, conservation and enhancement of financial, environmental and social capital (Ford, 2012). The study of Kaur Harjeet, (2011), examines the relationship between HR factors and perceived environmental performance of the employees in various business organizations and concluded that company should conduct energy audit, go paperless, recycle and get their employees motivated toward the greener eco effort and organisational sustainability. Green HR is a tool which can be help to reduce the costs. If the workforce is eco savvy the organization will have more inspired decision making to less stressed budget and improved employee retention (Paallavi, 2012). Green HR practices are likely to improve employee well being in the workplace not just improving the working environment and satisfying the need of an increasingly environmental aware workforce but also to contributes positively to both employee as well as organizational performance to achieve the goal of organisational sustainability (Gill Mandip, 2012). GHRM is an innovative approach to demonstrate how the organization manage sustainable development by initiating social conscience and green sense of responsibility as a part of organizational operation. Business organization today well about the green HR concept that has been put forward to help them to keep the environment green but still few business organisation are not able to put in to practices in different functional area of HRM (Popli, 2014). Green HRM is the use of the HRM policies to promote the sustainable use of the resources within the organization and more usually help the reason of environment sustainability (Rani & Mishra , 2014). HR plays vital role in management of people which is regarded as asset of an organization. The responsibility of the present HR managers is to incorporate the Green HR Philosophy in organizational objectives. Indeed many business organizations are adopting an integrated approach to implement Environmental Management programs .But achieving this integration of HR and Environment Management Involves importantly changing the approach of some employees themselves to green issues.

REVIEW OF LITERATURE

The very term 'sustainable HRM' has been used for more than a decade. The literature is piecemeal, diverse, and fraught with difficulties (Ehnert, 2009). There is no one precise definition of the term and it has been used in a variety of ways. The writings on sustainable HRM differ in terms of the emphasis given to particular internal and external outcomes. It has been used to refer to social and human outcomes which contribute to the continuation of the organization in the long-term. That is to a sustainable organization. It has also been used to refer to HRM activities which enhance positive environmental outcomes Green HRM (GHRM) and positive social and human outcomes for their own sake, rather than just as mediating factors between financial outcomes and strategy (Mazur, 2016). As with the terms HRM, SHRM and sustainability, there are definitional issues with the term sustainable HRM. A number of terms have been used to link sustainability and HRM activities (Kramar, 2014). These include sustainable work systems (SWSs), HR sustainability, sustainable management of HRs, sustainable leadership and sustainable HRM (Ehnert 2009). In addition, the term sustainable organization has also been used. They all recognize the impact HR outcomes have on the survival and success of the organization. A common feature of the writings on sustainable HRM is that HRM practices contribute to the development of the human and social capital within the organization. In addition, some of these writings also acknowledge that there is a growing concern for the impact of HRM policies on externalities, such as the environment and social and human aspects of society. Some of this literature also acknowledged that HRM practices will influence the extent to which people are attracted to work for an organization or to purchase its products and services. The literature on sustainable HRM can be categorized into three groups (Ehnert 2009). A common feature of all of these groups in an understanding that sustainability refers to long-term and durable outcomes.

RELATIONSHIP OF GREEN HRM WITH ORGANIZATIONAL SUSTAINABILITY

Adopting organization sustainability with specific employee job roles and descriptions enables organization to bring environmental strategies into daily development and production in such a way in which waste are reduce and managed, and all employees know how their own task and functions can promotes sustainable use of resources to promotes organizational sustainability (Van Veslor and Quinn 2012). Performance accountability forces organizations to monitor high–level standards set by top managements and provide sustainable working process and standards for each goal of the business organization (Crawford & Scaletta, 2005). In a study of Mesmer-Magnus et al. (2012) discussed that business organization can reach organizational sustainability and environmental sustainability simultaneously when all sides (individual, employees, organization) are committed. Hansla et al. (2008) believed that awareness about environmental issues and individuals' attitudes and behaviors has play an important role for pro-environmental actions and lack of awareness brings barriers and difficulties towards pro-environmental attitudes and behaviors for organizational sustainability. Hines et al (1987) represented that between Knowledge and high level awareness about environmental knowledge has more significant role for leading to pro environmental attitudes and behavior and helps employees setting their behavioral prioritize based on their effectiveness and environmental values to the organization.

Riketta and Dick (2005) discussed about an important predictor for employee commitment to organization. If the turnover is low then the organizational commitment is high, employee do not search for new job and leave the organization with low turnover, company will not invest on training to train new employees since existing employees will coach them and preserve and conserve the organization's culture, values, and competitiveness (Saks et al., 2007). In sum, it is obvious that employees are committed to their organization they will take actions for organizational sustainability initiatives (corporate social responsibility initiatives, environmental sustainability programs, etc.) to enhance organizational reputation and performances (Rettab et al., 2009). There are different levels of engagement to environment sustainability (D'Mello et al., 2011). Dimensions can be used in order to define the degree of engagement: (1) environmental sustainability initiatives, and (2) incorporation of environmental sustainability goals with organization's goals, operations, and functions as an explicit recognition. When employees perceive the organizational commitment to the sustainability, organizational commitment will increase. Also, it can shape trust to management (McWilliams & Siegel, 2001; D'Mello et al., 2011; Mesmer- Magnus et al., 2012). Commitment is the bond employee experiences with their organization broadly speaking employee who are committed to their organization generally feel that they fit in and feel they understand the environmental goal of the organization. The added value of such employee is that they tend to be a more determined in their work show relatively high productivity and are more proactive in offering their support toward organizational sustainability. With employees committed to the organization, which are passionate about their job, the bridge connecting organizational sustainability to environmental sustainability (Mesmer-Magnus et al., 2012).

THE HOLISTIC MODEL OF SUSTAINABLE HRM

Another important attempt to explain the complexity of the concept of sustainable Human Resource Management can be found in the research by De Prins (2011). He developed this model with the observations that the sustainable HRM focuses on optimally utilizing and respecting human workforces within an organization. The model further states that such optimal utilization and respect for human resources are built between organization's strategic policies and it's environment. The study has distinguished four approaches to the concept. The first approach has been identified as 'Psychological' and the second, third and fourth approaches have been identified as 'sociological', 'green' and 'strategic', respectively in mind that the stipulations of the related policies are still being reviewed.

The Holistic Model of Sustainable HRM Framework	
Psychological approach Focus what drives and characteristics	**Sociological approach** Explicit connection between the interests of the Employer, Employee and Society
Strategic approach HR's contribution to the long-term focus of the organization as a whole	**Green approach** Employee society Employee management and the planet

According to the sociological approach, HRM is considered as 'social practice'. This particular approach of administration has its focal point at continuity. The interests and involvement of three components-the employee, the employer, and the society are clearly interconnected. Of significant importance are such ideas as: engagement policies, health policies and societal themes like diversity, age-conscious and family friendly personnel policies. The psychological approach focuses on the employee, precisely on what subject matters he or she recognizes as relevant. As stated by De Prins, if people are at the centre of attention of a sustainable competitive advantage, then the knowledge and fostering of what drives and characteristics them is of ultimate significance. The strategic approach on sustainable HRM examines how sustainable HRM and strategic HRM are connected. The attention is aimed on the influence that sustainable HRM has on conventional HR spheres: recruitment and selection, employee turnover, appraisal and employability aspects of an organization. Ehnert's definition of sustainable HRM to describe this approach indicates that HR aims at achieving organizational goals, which entail more than benefits only. Typical themes are: the belief in humans as a sustainable competitive advantage, social achievements and the sustainable management of HR sources. Nevertheless, one should bear in mind that the stipulations of the related policies are still being reviewed.

Green HRM approach concerns the approaches of the employees and management to the planet component of the triple bottom line-which HRM aspects can help to make the organization 'green'. What is more, how does the 'green character' affect the employer's attractiveness and branding. Relevant themes are mentioning green behavior as a competence, training in sustainability awareness, stimulating environmentally conscious behaviours and green employer branding. Conclusion During the last two decades, a scholarly interest can be witnessed. The present paper attempts to discuss the ways, employees can be managed in the sustainable ways through available literature. As a result of a literature review, two models of sustainable HRM were discussed. The 'Three Pillars of Sustainable HRM' is the first model to emerge. It is narrower in scope than the second approach. It can be successfully applied in the study of national organizations and that is how it was first applied. The second model is broader in its nature, that is why it is called holistic. Because of its characteristics, it can be successfully used in the Research concerning the use of sustainable HRM in multinational and global companies. It seems, therefore, that the two models presented in this paper can be applied in analysis, however the choice of a model for study should depend on the purpose of the study and the size of the audited company. Nevertheless, the sustainable HRM theory seems fragmentary and incoherent, while the practical dimension requires developing. Sociology, systems science and public health are among the numerous fields that may offer substantial potential for academic regarding sustainable HRM.

CONCLUSION

It has been found that organization have been implementing a number of Green HRM initiatives to promotes sustainability. However, more serious approach needs to be taken to implement them, communicate them to the employees, and encourage employees and all stakeholders to adhere to such green initiatives, and reward them adequately. Also it is imperative to include the green HRM policy into various HRM function participation in green HR initiatives. The study focus on more eco-friendly working environment for the employees and the company stakeholders. The employers and practitioners can establish the usefulness of linking employee involvement and participation in environmental management programmes to promote organizational sustainability with a specific focus on waste management recycling, creating green products. Unions and employees can help employers to adopt Green HRM policies and practices that help safeguard and enhance worker health, safety, and well-being. The paper focuses on the involvement of HR in green initiative activities and sustainability. The study indicates that HR is involved in engaging the employees for implementing the Green HR initiatives. It creates a sense of morale and loyalty among the employees towards the organization sustainability which has also helped in employee recruitment and retention. When organization can reduce their cost and increase their profitability via green HRM. The future of Green HRM appears more promising in near future with the involvement of employees, management and stakeholders.

REFERENCES

Bansal, P. Hunter, T. (2003). Strategic Explanation for the Early Adoption. *Journal of Business Ethics 46*(3), 289-299.

Boudreau, J. & Ramstad, P. (2005). Talent ship, talent segmentation, and sustainability: A new HR decision science paradigm for a new strategy definition. *Human Resource Management, 44(2)*, 129–136.

Callenbach, E. Capra, F. Goldman, L. Lutz, R., & Marburg, S. (1993). *Eco-Management: The Elmwood Guide to Ecological Auditing and Sustainable Business,* Berretta-Koehle San Francisco, CA.

Carter, C. & Dressner, M. (2001). Purchasing role in environment management: Cross Functional development of Grounded theory. *Journal of supply chain management,* 12-27.

Cherian, J. & Jacob, J. (2012). A study of Green HR practices and its effective implementation in the organization: A review. *International Journal of Business and Management, 7*, 25–33.

Colbert, B. & Kurucz, E. (2007). *Three conceptions of triple bottom line business sustainability and the role for HRM,* Human Resource Planning.

Clarke, E. (2006), Power Brokers, People Management, 40-42.

Crawford, D. & Scaletta, T. (2005), the balanced scorecard and corporate social responsibility: *Aligning values for profit. CMA Management, 79*(6) 20-27.

Chesbrough, H. W. (2003). *Open Innovation: The New Imperative for Creating and Profiting from Technology,* Harvard Business School Press, Boston.

Cook, J. & Seith, B.J. (1992). Designs an effective environmental EMS training program. *Journal of Environmental Regulation, 2*(1), 53-62.

De Prins, P. (2011). DUURZAAM HRM: synthetische academische introductie, cited by: Rompa I., *Explorative Research on Sustainable Human Resource Management.* http://www. innovatiefinwerk.nl/ sites/... (accessed on September 15, 2018).

D'Mello, S., Wiernik, B.M., Ones. D. S., & Dilchert, S. (2011). 'The Relationship between Educational Level, Income, and Environmentalism: A Meta- Analysis. *Conference for Industrial and organizational Psychology, Chicago, Illinois.*

Ditz, D., Ranganathan, J., & Banks, R.D. (1995). *Green Ledgers: Case Studies in Corporate Environmental Accounting.* World Resources Institute, Washington.

Drucker, P. (1988). The coming of the new organization. *Harvard Business Review, 66* (1), 45–53.

Ford (2012). http://corporate.ford.com/doc/corpgov_sustainability_committee_charter.pdf.

Govindarajulu, N., & Daily, B.F. (2004). Motivating Employees for Environmental improvement. *Industrial Management and Data Systems, 104*, 364-372.

Gill, M. (2011). Green HRM: People Management Commitment to Environment Sustainability. *Research Journal of recent Science*, 244 -254.

Hansla, A., Gamble, A., Juliusson, A., & Garling, T. (2008). The relationships between awareness of consequences, environmental concern, and value orientations. *Journal of Environmental Psychology, 28* (1), 1-9.

Hirlikar, N. N. (n.d.) Green HRM Requirement of 21st Century. *International Multidisciplinary E – Journal – An International Peer Revived Preferred Journal*, 2277 – 4262.

Hines, J. M., Hungerford, H. R., & Tomera, A. N. (1987). Analysis and synthesis of research on responsible environmental behavior: A meta-analysis. *The Journal of environmental education*, 18 (2) 1-8.

Jackson, S. E., Renwick, D. W. S., Jabbour C. J. C., & Camen, M. M. (2011). State-of-the-Art and Future Directions for Green Human Resource Management: Introduction to the Special Issue. *German Journal of Research in Human Resource Management, 25*(2), 99-116.

Jain, G. (2015). Business ethics & the environment : Evolution of "Green HRM".

Kaur (2011). Impact Human Resource Factors on perceived Environmental performance: An empirical Analysis of a sample of ISO) EMS companies in Malaysia. *journal of sustainable development, 4*(1), 211-223.

Kramar, R. (2014). Beyond strategic human resource management: is sustainable human resource management the next approach? The International Journal of Human Resource Management, 25(8) : 1069-1089.

Kumari, P. (2012). Green HRM- Issue and Challenges. *GRA – Global Research Analysis*. 1(5).

Lee, K.H. (2009). Why and how to adopt green management into business organizations: the case Study of Korean SMEs in manufacturing industry'. *Management Decision, 47* (7) 1101–1121.

Malt, B. (2008). Speaking Green, *HR magazine*. Mesmer-Magnus, J., Viswesvaran, C., & Wiernik,

Mazur, B. (2016). Green human resource management, In B.Kozuch & K. Sien-KiewiczMalyjurek (eds.), Faces of contemporary management, Krakow; Jagiellonian University Press, pp. 11-22.

Riketta, M. & Dick, R. V. (2005). Foci of attachment in organizations: A met analytic comparison of the strength and correlates of workgroup versus organizational identification and commitment. *Journal of Vocational Behavior*, 67 (3) 490-510.

CHAPTER 11

Transforming Attitudes Towards Entrepreneurial Skills in B.Ed Students through Edupreneur Project Training

DR. GEETA PATHAK & DR. ARTI BHATNAGAR

Associate Professors, Guru RamDass College of Education, Delhi, India

ABSTRACT

This study investigates the impact of Edupreneur Project Training on the entrepreneurial self-efficacy, risk-taking behavior, and innovation behavior of B.Ed students. Utilizing three standardized scales, namely the Entrepreneurial Self-Efficacy Scale, Jackson Personality Inventory Risk-Taking Scale, and Individual Innovation Behavior Scale, data were collected from 73 participants before and after the training intervention. The analysis revealed significant improvements across all dimensions, with post-test scores indicating heightened levels of entrepreneurial self-efficacy, risk-taking propensity, and innovation behavior compared to pre-test scores. Specifically, the mean post-test scores for risk-taking behavior and innovation behavior increased substantially, supported by low variability and high T-ratios, signifying statistical significance. Furthermore, the analysis of entrepreneurial self-efficacy demonstrated a prevalent positive self-assessment among participants, suggesting a robust belief in their entrepreneurial abilities. These findings underscore the efficacy of Edupreneur Project Training in cultivating essential entrepreneurial skills and attitudes among future educators. The study's implications highlight the importance of integrating entrepreneurship education into teacher training programs to empower educators with the necessary tools to foster innovation, creativity, and entrepreneurial thinking among students. Ultimately, investing in entrepreneurial education for educators has the potential to drive societal change, economic growth, and the cultivation of a new generation of entrepreneurial leaders poised to tackle the challenges of the future.

Keywords: *Entrepreneurial Skills, Edupreneur Project Training, B.Ed. Students, Attitude Change, Educational Innovation, Entrepreneurial Self-efficacy*

INTRODUCTION

The fast changing nature of education has made it clear that one of the most important ways to get future teachers ready for the changing needs of the twenty-first century is to incorporate entrepreneurial skills into teacher preparation programmes.

The importance of entrepreneurship in teacher education stems from its capacity to support creative teaching methods, develop educators' adaptability and resilience, and ultimately improve student learning results. Through the cultivation of entrepreneurial skills like innovation, problem-solving, and risk-taking, teacher education programmes can equip aspiring teachers with the necessary tools to effectively address a variety of classroom requirements and interact with evolving educational technologies.

But there are certain difficulties in incorporating business into teacher preparation.

- Curriculum change is hampered by traditional educational approaches, which frequently place a higher priority on material knowledge than on entrepreneurial abilities.
- Further impeding success in this area include faculty reluctance, scarce resources, and a lack of knowledge about the importance of entrepreneurship in education.

To address these challenges, strategic initiatives are essential. These may include curriculum redesign to embed entrepreneurial principles across subject areas, professional development programs to train faculty in entrepreneurial pedagogy, and collaborations with industry partners to provide real-world learning opportunities for pre-service teachers. A disruptive paradigm in teacher education is emerging with the goal of giving educators the tools and perspective they need to successfully negotiate the intricacies of today's educational environments: "Edupreneurship"

The term "Edupreneurship" refers to the adoption of entrepreneurial ideas and methods by teachers in order to foster innovation, enhance student learning, and add value to respective institutions of higher learning. The following are five doable tactics for edupreneurship in the teaching field:

- **Innovation in Curriculum:** To better serve the varied needs of their students, entrepreneurial educators constantly look for new and exciting ways to innovate and improve the curriculum. To increase the relevance and impact of learning, they create lesson plans that are interesting and combine technology in an efficient manner, as well as real-world applications.
- **Student-Centered Learning:** Edupreneurial teachers prioritize student-centered learning approaches that empower students to take ownership of their learning journey. They foster creativity, critical thinking, and problem-solving skills by designing projects, activities, and assessments that encourage active participation and collaboration.
- **Entrepreneurial Projects:** Integrate entrepreneurial projects into the classroom to provide students with hands-on experiences in entrepreneurship and business concepts. These projects can involve designing and marketing products, organizing events, or creating social enterprises, fostering entrepreneurship skills and mindset among students.
- **Professional Development:** Engage in continuous professional development to enhance entrepreneurial competencies and pedagogical skills. Edupreneurial teachers attend workshops, conferences, and online courses to stay updated on the latest trends in education, technology, and entrepreneurship, enabling them to innovate and adapt to changing educational landscapes.
- **Community Engagement:** Collaborate with local communities, businesses, and organizations to create authentic learning experiences for students. Edupreneurial teachers leverage community resources, invite guest speakers, and organize field trips to expose students to real-world problems and opportunities, fostering entrepreneurial thinking and civic engagement.

By embracing Edupreneurship, teachers can transform their classrooms into dynamic learning environments that inspire creativity, innovation, and lifelong learning among students. Through curriculum innovation, student-centered learning, entrepreneurial projects, professional development, and community engagement, Edupreneurial teachers can empower the next generation of innovators, leaders, and change makers to thrive in the 21st-century economy. Entrepreneurship in teacher education holds immense promise for driving educational innovation and preparing educators to thrive in an ever-evolving educational landscape. By overcoming challenges through strategic interventions, stakeholders can harness the transformative potential of entrepreneurship to nurture a new generation of empowered and innovative educators.

LITERATURE REVIEW

Entrepreneurship is increasingly recognized as a critical component of education, essential for fostering innovation and economic development (Kuratko, 2005). As future educators, B.Ed. students play a pivotal role in imparting entrepreneurial skills to the next generation. However, many teacher education programs do not adequately prepare students for this role. This study investigates the impact of Edupreneur Project Training (EPT) on transforming B.Ed students' attitudes towards entrepreneurial skills.

Kuratko (2005) traced the emergence and development of entrepreneurship education, highlighting its growth in higher education institutions worldwide. Fayolle and Gailly (2008) examined teaching models and learning processes in entrepreneurship education, proposing a framework for designing and evaluating entrepreneurship education programs. Rae (2010) investigated the role of universities in promoting entrepreneurship education, emphasizing the need for curricular adaptations to foster entrepreneurial skills. Bae et al. (2014) conducted a meta-analysis revealing a positive correlation between entrepreneurship education and entrepreneurial intentions. Nabi et al. (2017) conducted a systematic review of the literature on the impact of entrepreneurship education in higher education, identifying the need for more longitudinal studies. Lackéus (2015) provided an overview of entrepreneurship education, emphasizing its benefits in developing entrepreneurial mindsets among students. Neck and Corbett (2018) reviewed the scholarship of teaching and learning entrepreneurship, highlighting the need for more rigorous research. Rauch and Hulsink (2015) found that entrepreneurship education influences entrepreneurial behavior by shaping intentions and self-efficacy. Sirelkhatim and Gangi (2015) identified key trends in entrepreneurship education, such as the emphasis on practical experiences and interactive teaching methods.

Rideout and Gray (2013) critiqued empirical literature on the effects of university-based entrepreneurship education, proposing guidelines for rigorous research. Fayolle et al. (2016) discussed the theoretical and methodological foundations of entrepreneurship education research, calling for a robust research agenda. Pittaway and Cope (2007) conducted a systematic review, identifying trends like the focus on entrepreneurial intentions and experiential learning methods. Liñán (2004) proposed an intention-based model, highlighting the importance of aligning entrepreneurship education with the target audience's needs. Mwasalwiba (2010) reviewed the objectives and teaching methods of entrepreneurship education, proposing a framework for evaluating its effectiveness. Hattab (2014) found that entrepreneurship education positively influences entrepreneurial intentions by shaping attitudes and perceived behavioral control. Souitaris et al. (2007) showed that entrepreneurship programs raise entrepreneurial intentions by providing learning opportunities and resources. Fayolle and Gailly (2015) found that entrepreneurship education has a lasting impact on entrepreneurial attitudes and intentions.

Piperopoulos and Dimov (2015) investigated the relationship between entrepreneurship education, self-efficacy, and intentions, suggesting the need to develop both skills and self-efficacy.

Nabi et al. (2018) examined the impact of first-year entrepreneurship education on intentions, suggesting early introduction in the curriculum. Blenker et al. (2014) reviewed research methods in entrepreneurship education, proposing an integrative framework for future studies. Collectively, these studies underscore the significance of entrepreneurship education in shaping students' entrepreneurial skills and intentions. The reviewed literature emphasizes the need for well-designed entrepreneurship education programs that incorporate practical experiences, interactive methods, and contextual factors. These insights are particularly relevant for B.Ed programs, where preparing future educators to impart entrepreneurial skills is crucial. Integrating entrepreneurship education into teacher training can enhance teachers' effectiveness and adaptability, fostering a dynamic educational environment. This study aims to fill the research gap by examining the impact of Edupreneur Project Training on B.Ed students, contributing valuable insights for curriculum development and policy formulation.

JUSTIFICATION OF THE STUDY

In light of the significant influence educators exert on shaping the outlook and capabilities of future generations, imparting them with entrepreneurial competencies holds the potential to catalyze transformative shifts within the educational landscape, fostering a climate conducive to innovation and adaptability. The introduction of BED 210-Entrepreneurship in Education within the B.Ed. curriculum at GGSIPU in 2022 underscores a proactive step towards integrating entrepreneurial education into teacher training programs. In response to this initiative, there arises a critical imperative to evaluate the attitudes and perceptions of trainee teachers towards this novel course and its consequent impact. Thus, this study seeks to bridge existing research gaps by specifically examining the effects of entrepreneurial training on B.Ed. students, offering valuable insights essential for informing curriculum design and policy development initiatives aimed at nurturing entrepreneurial mindsets among educators.

Objectives of the Study

1. Primary Objective: To assess the effectiveness of Edupreneur Project Training in transforming attitudes towards entrepreneurial skills among B.Ed students.
2. Secondary Objectives:
   - To evaluate the impact of Edupreneur Project Training on students' entrepreneurial self-efficacy.
   - To examine changes in risk-taking behavior among B.Ed students following participation in the training program.
   - To assess the influence of Edupreneur Project Training on individual innovation behavior among B.Ed students.

Research Questions

1. What is the impact of Edupreneur Project Training on B.Ed students' entrepreneurial self-efficacy?

2. How does participation in Edupreneur Project Training affect B.Ed students' risk-taking behavior?
3. What changes occur in B.Ed students' individual innovation behavior after undergoing Edupreneur Project Training?

Methodology

A quasi-experimental design with pre-test and post-test assessments will be employed. The whole group will Analyzed for their Risk Taking & Innovation Behaviour Score Before the Intervention & then after the Treatment/ Intervention (Edupreneur Project Training), the Sample will be Analyzed on their Risk Taking, Innovation Behaviour & Entrepreneurial Self-efficacy Scores.

Sample

The study involved a sample of 73 trainee teachers enrolled in Bachelor of Education (B.Ed) programs in a Teacher Training College affiliated to GGSIPU, Delhi. The sample diversification of 74 participants is as follows:

Table 1: Sample Diversification

Sample Diversification	*Total (73 students)*
Male : Female	08 : 65
Graduates : Post-graduates	64 : 09

TOOLS OF THE STUDY

Three Standardized Tools were used to check the impact of Edupreneur Project Training on B.Ed Students. Two Tools were used as Pre & Post test where as the third tool was used as Post test to check the Efficacy & Impact of Edupreneur Project Training on B.Ed Students.

Tool I: Jackson Personality Inventory Risk-Taking Scale (Jackson, 1977)

The Jackson Personality Inventory Risk-Taking Scale (Jackson, 1977) is a psychological assessment tool designed to measure an individual's propensity for risk-taking behavior across various life domains. This scale aims to capture an individual's willingness to engage in risky activities, decisions, or behaviors that potentially involve uncertain outcomes. The scale typically consists of a series of statements or items that respondents rate on a Likert-type scale based on the degree to which they agree or disagree with each statement. Reliability of the RTS scale is demonstrated through high internal consistency, with Cronbach's alpha values typically reported above 0.76, indicating strong reliability. This robust reliability and validity make the ESE scale a valuable tool for both academic research and practical applications in entrepreneurship education and development programs.

This scale was used as pre test & post test to check the impact of Edupreneur Project Training on B.Ed students.

Tool II: Individual Innovation Behavior Scale (Scott & Bruce, 1994)

The Individual Innovation Behavior Scale (IIBS), developed by Scott and Bruce in 1994, is a psychometric tool designed to assess an individual's propensity for engaging in innovative behaviors within organizational settings. The scale aims to measure various dimensions of innovation behavior, including generating new ideas, implementing novel solutions, and championing change initiatives. It typically consists of a series of items or statements that respondents rate based on the frequency or extent to which they engage in specific innovative behaviors. Reliability of the ESE scale is demonstrated through high internal consistency, with Cronbach's alpha values typically reported above 0.80, indicating strong reliability. This robust reliability and validity make the ESE scale a valuable tool for both academic research and practical applications in entrepreneurship education and development programs.

This scale was used as pre test & post test to check the impact of Edupreneur Project Training on B.Ed students.

Tool III: Entrepreneurial Self-Efficacy Scale by Chen, Greene & Crick 1998

The Entrepreneurial Self-Efficacy (ESE) scale, developed by Chen, Greene, and Crick (1998), measures an individual's belief in their ability to perform tasks essential for entrepreneurship. It is a multidimensional construct that assesses various aspects of entrepreneurial tasks. The scale includes items that reflect the confidence in:

1. Opportunity Recognition: Identifying viable business opportunities.
2. Innovation: Developing new products or services.
3. Risk Management: Handling risks associated with entrepreneurial activities.
4. Resource Acquisition: Securing necessary financial, human, and informational resources.
5. People Management: Leading and managing a team effectively.

Reliability of the ESE scale is demonstrated through high internal consistency, with Cronbach's alpha values typically reported above 0.80, indicating strong reliability. This robust reliability and validity make the ESE scale a valuable tool for both academic research and practical applications in entrepreneurship education and development programs.

This Scale was used Post Test to check the impact of Edupreneur Project Training on B.Ed students.

DATA ANALYSIS

Data Analysis of Tool I: RTS Scale

Table 2: Data Analysis of RTS Scale Before and After the Edupreneur Project Training of B.Ed Students (N = 73)

Dimension	*Pre Test*		*Post Test*		*T Ratio*
	Mean	*SD*	*Mean*	*SD*	
Risk Taking Ability	2.99	0.81	3.97	0.73	16.64

The analysis reveals a significant improvement in risk-taking behavior from the pre-test to the post-test. The mean pre-test score was 2.99, while the mean post-test score increased to 3.97. This indicates that, on average, students exhibited a higher level of risk-taking behavior in the post-test compared to the pre-test. The standard deviation of the post-test scores (0.73) suggests relatively low variability in risk-taking tendencies among the students after the intervention. The calculated T-ratio of 16.64 indicates a statistically significant difference between the mean post-test and pre-test scores ($p < 0.05$). This suggests that the improvement in risk-taking behavior from the pre-test to the post-test is unlikely to have occurred by chance alone, providing evidence that the intervention or treatment implemented between the pre-test and post-test assessments effectively enhanced students' risk-taking propensity.

Data Analysis of Tool II: IIB Scale

Table 3: Data Analysis of IIV Scale Before & After the Edupreneur Project Training of B.Ed Students (N = 73)

Dimension	*Pre Test*		*Post Test*		*T Ratio*
	Mean	*SD*	*Mean*	*SD*	
Innovation Behaviour Score	2.96	0.79	4.02	0.75	19.47

The analysis indicates a significant improvement in innovation behavior from the pre-test to the post-test. The mean pre test score was 2.96, while the mean post-test score increased to 4.02. This suggests that, on average, students exhibited a higher level of innovation behavior in the post-test compared to the pre-test. The standard deviation of the post-test scores (0.75) suggests relatively low variability in innovation behavior among the students after the intervention. The calculated T-ratio of 19.47 indicates a statistically significant difference between the mean post-test and pre-test scores ($p < 0.05$). This suggests that the improvement in innovation behavior from the pre-test to the post-test is highly unlikely to have occurred by chance alone, providing strong evidence that the intervention or treatment implemented between the pre-test and post-test assessments effectively enhanced students' innovation behavior.

Data Analysis of Tool III:ESE Scale

Table 4: Data Analysis of ESE Scale after the Edupreneur Project Training of B.Ed Students (N=73)

Dimension	*Mean*	*SD*	*T-ratio*
Opportunity Recognition	3.88	0.89	8.85
Innovation	3.82	0.87	8.17
Risk Management	3.77	0.93	6.92
Resource Acquisition	3.74	0.92	6.44
People Management	3.71	0.90	6.28

The analysis of the dataset on Entrepreneurial Self-Efficacy (ESE) scale for 73 students reveals notable insights into their perceived abilities across different dimensions of entrepreneurship. The mean scores for each dimension, ranging from 3.71 to 3.88, all exceed the midpoint of the scale (3), indicating a general positive self-assessment among the students regarding their entrepreneurial competencies. This suggests that, on average, the students believe in their capabilities to identify opportunities, innovate, manage risks, acquire resources, and lead teams effectively in entrepreneurial endeavors. Additionally, the relatively low standard deviations (ranging from 0.87 to 0.93) imply a consistent level of confidence among the students, with limited variability in their self-efficacy ratings across the sample.

Moreover, the T-ratios, which assess whether the means significantly differ from the midpoint of the scale, further elucidate the strength of the students' entrepreneurial self-efficacy. With T-ratios ranging from 6.28 to 8.85, all dimensions exhibit significantly higher mean scores compared to the midpoint, indicating a robust belief in their entrepreneurial abilities. This suggests a collective confidence among the students in their capacity to recognize opportunities, innovate solutions, manage risks, secure resources, and lead teams, aligning well with the foundational principles of entrepreneurship. Overall, the analysis underscores a positive self-assessment and a strong sense of self-efficacy among the students, reflecting a promising outlook for their potential engagement and success in entrepreneurial pursuits.

CONCLUSION

The analysis of the dataset reveals a significant improvement in innovation behavior among B.Ed students following participation in the Edupreneur Project Training. The substantial increase in mean post-test scores compared to pre-test scores indicates a noteworthy enhancement in students' inclination towards innovative behaviors. The low variability in post-test scores further strengthens the evidence, suggesting a consistent positive shift across the sample. With a high T-ratio indicating statistical significance, it is evident that the intervention effectively contributed to enhancing students' innovation behavior, demonstrating the program's efficacy in fostering a culture of innovation among future educators. Additionally, the examination of students' entrepreneurial self-efficacy underscores a prevalent positive self-assessment among participants regarding their entrepreneurial competencies. The consistently high mean scores across different dimensions of entrepreneurship, coupled with low standard deviations, highlight a collective confidence among students in their ability to excel in various entrepreneurial tasks. This robust belief in their capabilities bodes well for their future engagement and success in entrepreneurial endeavors, emphasizing the importance of fostering entrepreneurial mindset and skills within educational contexts.

The findings of this study hold significant implications for educational institutions and policymakers aiming to promote entrepreneurship and innovation among B.Ed students. By demonstrating the effectiveness of targeted interventions such as the Edupreneur Project Training in enhancing students' innovation behavior and entrepreneurial self-efficacy, the study highlights the importance of integrating entrepreneurship education into teacher training programs. Implementing similar initiatives can empower future educators to not only excel in their teaching roles but also inspire creativity, problem-solving, and entrepreneurial thinking among their students. Moreover, fostering a culture of innovation within educational institutions can contribute to addressing societal challenges, driving economic growth, and cultivating a new generation of entrepreneurial leaders equipped to navigate the complexities of the modern world. Thus, investing in entrepreneurial education for educators has the potential to yield long-term benefits for individuals, communities, and economies alike.

REFERENCES

Bae, T.J., Qian, S., Miao, C., & Fiet, J.O. (2014). The relationship between entrepreneurship education and entrepreneurial intentions: A meta-analytic review. *Entrepreneurship Theory and Practice*, 38(2), 217-254.

Barbosa, S. D., Gerhardt, M. W., & Kickul, J. R. (2007). The role of cognitive style and risk preference on entrepreneurial self-efficacy and entrepreneurial intentions. *Journal of Leadership & Organizational Studies*, 13(4), 86-104.

Blenker, P., Dreisler, P., & Kjeldsen, J. (2014). Entrepreneurship education–the new challenge facing the universities. *Education & Training.*

Chen, C. C., Greene, P. G., & Crick, A. (1998). Does entrepreneurial self-efficacy distinguish entrepreneurs from managers? *Journal of Business Venturing*, 13(4), 295-316.

Fayolle, A., & Gailly, B. (2008). From craft to science: Teaching models and learning processes in entrepreneurship education. *Journal of European Industrial Training*, 32(7), 569-593.

Fayolle, A., & Gailly, B. (2015). The impact of entrepreneurship education on entrepreneurial attitudes and intention: Hysteresis and persistence. *Journal of Small Business Management*, 53(1), 75-93.

Fayolle, A., Verzat, C., & Wapshott, R. (2016). In quest of legitimacy: The theoretical and methodological foundations of entrepreneurship education research. *International Small Business Journal*, 34(7), 895-904.

Hattab, H. W. (2014). Impact of entrepreneurship education on entrepreneurial intentions of university students in Egypt. *The Journal of Entrepreneurship*, 23(1), 1-18.

Jackson, D.N. (1977). Jackson Personality Inventory. Research Psychologists Press.

Kuratko, D.F. (2005). The emergence of entrepreneurship education: Development, trends, and challenges. *Entrepreneurship Theory and Practice*, 29(5), 577-598.

Kuratko, D.F. (2005). The emergence of entrepreneurship education: Development, trends, and challenges. *Entrepreneurship Theory and Practice*, 29(5), 577-597.

Lackéus, M. (2015). Entrepreneurship in education: What, why, when, how. OECD.

Liñán, F. (2004). Intention-based models of entrepreneurship education. *Piccolla Impresa/Small Business.*

Mwasalwiba, E.S. (2010). Entrepreneurship education: a review of its objectives, teaching methods, and impact indicators. *Education & Training*, 52(1), 20-47.

Nabi, G., Liñán, F., Fayolle, A., Krueger, N., & Walmsley, A. (2017). The impact of entrepreneurship education in higher education: A systematic review and research agenda. *Academy of Management Learning & Education*, 16(2), 277-299.

Nabi, G., Walmsley, A., Liñán, F., Akhtar, I., & Neame, C. (2018). Does entrepreneurship education in the first year of higher education develop entrepreneurial intentions? The role of learning and inspiration. *Studies in Higher Education*, 43(3), 452-467.

Neck, H.M., & Corbett, A.C. (2018). The scholarship of teaching and learning entrepreneurship. *Entrepreneurship Education and Pedagogy*, 1(1), 8-41.

OECD (2010). The OECD Innovation Strategy: Getting a Head Start on Tomorrow. OECD Publishing.

Piperopoulos, P., & Dimov, D. (2015). Burst bubbles or build steam? Entrepreneurship education, entrepreneurial self-efficacy, and entrepreneurial intentions. *Journal of Small Business Management*, 53(4), 970-985.

Pittaway, L., & Cope, J. (2007). Entrepreneurship education: A systematic review of the evidence. *International Small Business Journal*, 25(5), 479-510.

Rae, D. (2010). Universities and enterprise education: responding to the challenges of the new era. *Journal of Small Business and Enterprise Development*, 17(4), 591-606.

Rauch, A., & Hulsink, W. (2015). Putting entrepreneurship education where the intention to act lies: An investigation into the impact of entrepreneurship education on entrepreneurial behavior. *Academy of Management Learning & Education*, 14(2), 187–204.

Rideout, E.C., & Gray, D.O. (2013). Does entrepreneurship education really work? A review and methodological critique of the empirical literature on the effects of university-based entrepreneurship education. *Journal of Small Business Management*, 51(3), 329–351.

Scott, S.G., & Bruce, R.A. (1994). Determinants of innovative behavior: A path model of individual innovation in the workplace. *Academy of Management Journal*, 37(3), 580–607.

Sirelkhatim, F., & Gangi, Y. (2015). Entrepreneurship education: A systematic literature review of curricula contents and teaching methods. *Cogent Business & Management*, 2(1), 1052034.

Souitaris, V., Zerbinati, S., & Al-Laham, A. (2007). Do entrepreneurship programmes raise entrepreneurial intention of science and engineering students? The effect of learning, inspiration and resources. *Journal of Business Venturing*, 22(4), 566–591.

Stewart, W.H., & Roth, P.L. (2001). Risk propensity differences between entrepreneurs and managers: A meta-analytic review. *Journal of Applied Psychology*, 86(1), 145–153.

West, M.A. (2002). Sparkling fountains or stagnant ponds: An integrative model of creativity and innovation implementation in work groups. *Applied Psychology: An International Review*, 51(3), 355–387.

CHAPTER 12

Students' Attitudes Towards Community Service: Implications for Effective Implementation

Dr. Manvinder Kaur

Assistant Professor, Guru Ram Dass College of Education, GGSIPU, Delhi

Abstract

Education serves as a powerful tool for social change. Its primary responsibility is to equip students for active, constructive roles within society. This objective is realized through community service practices. As the profession of teaching too is an act of welfare for the society, hence community service finds more relevance for teacher training programme. This study investigates the attitudes of B.Ed. Students towards community service and explores how these attitudes impact the effectiveness of community service activities. A mixed-method approach, combining survey and focus group discussions, was employed to gather data from a sample of 100 B.Ed. Students. The findings reveal that while most students recognize the value of community service, their willingness to participate is influenced by several factors, including perceived benefits, mandatory participation, and support from educational institutions. The study concludes with recommendations for designing and implementing community service activities that enhance student engagement and foster a positive attitude towards community involvement. NEP 2020 too envisions crucial goal of fostering social responsibility and community engagement in HEIs. Hence It is envisaged that these findings will help stakeholders give space to community service, engagement and participation in the curricula.

Keywords: *Community Service, Community Engagement, Teacher Training, NEP 2020, Attitudes, Citizenship, Social Skills*

INTRODUCTION

Community

Common traits among its members—such as location, class, race, culture, language, customs, and laws—define a community. Certain communities exhibit homogeneity, whilst others display heterogeneity. Enhancing the quantity and quality of educational outcomes is thought to be possible through community participation in education. Every member of a community has a distinct role that is frequently in line with the group's overarching objectives.

Community Service

It is working with community and not working for community. It means one has to learn how to work with community members. Engaging in community service allows individuals to socialize and connect with the community around them. It provides a unique opportunity to learn new skills and gain insights from community members. Through this interaction, one could explore diverse lifestyles and cultural practices. Community service could be undertaken individually or as part of a group, fostering collaboration and mutual growth. Both students and the community benefit from this exchange, as students gain valuable experience and communities receive support and fresh perspectives.

Community service has become an integral component of educational programs (higher and school education) aimed at fostering civic engagement and social responsibility among students. As teacher trainee programs like B.Ed. increasingly incorporate community service into their curricula, understanding students' attitudes towards these activities is crucial for developing effective implementation strategies. Students are able to apply their academic knowledge in practical contexts, gain important social and emotional skills, and strengthen their sense of civic responsibility and social duty by participating in community service. This study aims to explore B.Ed. students' perceptions of community service, identify factors influencing their willingness to engage, and provide actionable insights for educators and policymakers.

Importance of Community Service Activities

Engaging in community service enables students to understand local lifestyles and socio-economic challenges. Through hands-on experiences, they develop insights into social issues and foster empathy. Interaction with the community facilitates problem identification and mutual assistance. Organizing events and surveys raises awareness and initiates discussions. Participation in activities like cleaning and tree planting enhances community bonds. This involvement strengthens the relationship between college and community, promoting collaborative efforts for positive change. Overall, community service empowers students to address societal issues effectively while contributing to the well-being of their local communities.

Community Service as Envisioned by NEP 2020

Community service has emerged as a vital component of education systems worldwide, including in India. The National Education Policy (NEP) 2020, introduced by the Government of India, underscores the significance of integrating community service into the educational framework to foster holistic development and civic responsibility among students. This policy shift aims to equip students not only with academic knowledge but also with the skills, values, and attitudes necessary for active and meaningful participation in society.

- "11.8. *Towards the attainment of such a holistic and multidisciplinary education, the flexible and innovative curricula of all HEIs shall include credit-based courses and projects in the areas of community engagement and service, environmental education.*" (NEP 2020, p. 38)
- "12.9. *Students are the prime stakeholders in the education system. Vibrant campus life is essential for high-quality teaching-learning processes. Towards this end, students will be given plenty of opportunities for participation in sports, culture/arts clubs, eco-clubs, activity clubs, community service projects, etc.*" (NEP 2020, p. 41)
- "15.6. *HEIs offering teacher education programmes will ensure the availability of a range of experts in education and related disciplines as well as specialized subjects. Each higher education institution will have a network of government and private schools to work closely with, where potential teachers will student-teach along with participating in other activities such as community service, adult and vocational education, etc.*" (NEP 2020, p. 44)

NEP 2020 highlights the value of experiential learning and the acquisition of essential life skills, as can be seen above. One of the most important components in reaching these educational objectives is community service. The policy proposes a number of approaches to incorporate community service into the curriculum:

(a) **Curriculum Integration:** Community service should be incorporated into the curriculum to give students the chance to apply what they have learned in real-world settings. The goal of this integration is to increase the relevance and impact of learning.
(b) **Skill Development:** Students could acquire vital life skills including problem-solving, cooperation, leadership, and communication through community service. These abilities are developed via practical, community-based experiences and are essential for both personal and professional success.
(c) **Civic Responsibilities:** NEP 2020 seeks to instil in students a feeling of social and civic responsibility. Students could gain an understanding of the value of tackling social concerns and making a contribution to society through community service.
(d) **Holistic Development:** The policy promotes students' cognitive, emotional, social, and ethical growth in a holistic manner. Participating in community service projects offers a platform for this holistic development.
(e) **Community Engagement:** In line with NEP 2020, educational institutions are urged to establish alliances with local governments, NGOs, and community organisations. These partnerships have the potential to increase the reach and influence of community service initiatives.

The NEP 2020 envisions a transformative approach to education, emphasizing experiential learning, critical thinking, and ethical awareness. One of its key objectives is to nurture socially conscious individuals who are capable of contributing positively to their communities. By integrating community service into the curriculum, NEP 2020 seeks to bridge the gap between theoretical learning and practical application, enabling students to address real-world challenges and develop a sense of social responsibility.

LITERATURE REVIEW

Research indicates that community service has numerous benefits for students, including the development of social skills, enhanced academic performance, and increased awareness of societal issues (Billig, 2000; Youniss & Yates, 1997). Community service helps students develop a sense of empathy and responsibility towards their communities, fostering a spirit of civic-mindedness that could last a lifetime. These experiences are particularly valuable in cultivating leadership skills and a sense of agency, empowering students to effect positive change in their communities. However, students' attitudes towards community service could vary significantly based on personal, social, and institutional factors (Metz & Youniss, 2003). Previous studies have shown that mandatory community service requirements could sometimes lead to negative perceptions, reducing the intrinsic motivation to participate (Stukas, Snyder, & Clary, 1999). Students who are required to engage in community service as a compulsory component of their education may perceive it as a burden or an obligation, rather than an opportunity for personal growth and community contribution. This could lead to feelings of resentment and disengagement, ultimately undermining the goals of community service activities.

Conversely, voluntary participation and support from teachers and peers have been linked to more positive attitudes (Serow, 1991). When students are given the choice to participate in community service, they are more likely to view it as a meaningful and fulfilling activity. Positive reinforcement from educators and the presence of a supportive peer network could significantly enhance students' motivation and engagement. Additionally, the alignment of community service activities with students' interests and passions is crucial in fostering a positive attitude towards participation. Given the potential of community service to positively impact both students and communities, it is essential to explore how students perceive these activities and what motivates or discourages their participation. This understanding could inform the design and implementation of community service activities that not only achieve educational objectives but also inspire a lifelong commitment to civic engagement and social responsibility.

OBJECTIVES OF THE STUDY

1. To assess students' perceptions of the importance and benefits of community service
2. To identify motivations and barriers influencing students' willingness to engage in community service.
3. To investigate the students' attitudes towards mandatory versus voluntary community service.
4. To provide recommendations for designing community service activities.

METHODOLOGY

Population

All the students (teacher trainees) enrolled in B.Ed. program run by the 26 colleges affiliated with Guru Gobind Singh Indraprastha University, Delhi.

Sample

A total of 100 B.Ed. students from second year from one B.Ed. College in East Delhi participated in this study. The sample included a diverse group of students from various socio-economic backgrounds, ensuring a comprehensive understanding of the attitudes across different demographics.

Research Tools

1. **Survey:** A questionnaire was designed to assess students' attitudes towards community service, including perceived benefits, willingness to participate, and factors influencing their attitudes. The survey included closed-ended items to capture a clear snapshot of their perspectives and experiences.
2. **Focus Group Discussions:** To gain deeper insights, focus group discussions were conducted with a subset of 30 students. These discussions aimed to explore students' experiences and perceptions in more detail, providing a qualitative complement to the quantitative survey data.

Procedure

The survey was administered to B.Ed. Students after they had participated in various community Service Activities in the two-year duration of their B.Ed. Program-Discussion, Awareness Rally, Cleanliness Drive, Nukkad Natak, Poster and Banner display, Theme based cultural programme, Survey, interview, action research, case study, participant observation, Interaction with community members. All these activities have them a real time experience and better understanding of Community services. Focus group discussions were held post college timings to ensure voluntary participation and to provide a relaxed and open environment for students to share their thoughts and experiences. Data were analysed using both quantitative and qualitative methods. Survey responses were statistically analysed to identify patterns and correlations, while thematic analysis was used for the focus group discussions to identify common themes and insights.

RESULTS

Survey Findings

Key factors influencing students' attitudes included:

(a) **Perceived Benefits:** 70% of students viewed community service positively, recognizing its benefits for personal development and societal impact. 78% of students believed that community service would enhance their college applications or job prospects.

(b) **Mandatory versus Voluntary Participation:** While 60% of students felt that mandatory community service was beneficial, 40% reported feeling resentful or burdened by the requirement.

(c) **Institutional Support:** 60% of students who received encouragement and support from teachers and school administrators were more positive attitudes towards community service.

(d) **Social Integration:** 85% of students felt feeling a greater sense of belonging and connection to their community through participation in community service activities.

(e) **Personal Growth:** 65% of students experienced increased self-confidence and personal growth as a result of their involvement in community service projects.

(f) **Skill Development:** 70% of students felt they acquired new skills, such as leadership, communication, and teamwork, through their engagement in community service.

(g) **Long term Commitment:** 70% of students expressed a desire to continue participating in community service even after completing their academic requirements, indicating a lasting commitment to social responsibility.

(h) **Impact on Academic Performance:** 68% of students felt engagement in community service activities increased their interest in academics also. Indicating that a well-designed community service activity balances the curricular load for the students.

Focus Group Insights

Focus group discussions revealed several themes. Some key responses from the participants are also mentioned.

(a) **Intrinsic Motivation:** Students who had prior positive experiences with community service were more intrinsically motivated to participate. They expressed a sense of fulfilment and personal growth, emphasizing the importance of meaningful and impactful activities.

- *"I've found that every time I engage in community service, I come away feeling more fulfilled and accomplished. It's like I'm making a real difference, and that drives me to want to do more."*
- *"For me, community service isn't just about fulfilling a requirement; it's about personal growth. The feeling you get from knowing you've positively impacted someone's life is indescribable. It's what keeps me coming back for more."*

b) **Peer Influence:** The presence of friends or peers in community service activities significantly increased students' willingness to engage. The social aspect of community service, including the opportunity to bond with peers and work collaboratively towards a common goal, was highlighted as a key motivator.

- *"Honestly, it's empowering to work together towards a shared objective, especially with familiar faces by my side."*
- *"Seeing my friends involved in community service activities inspired me to join in. It's like we're a team, working together to make a difference. Plus, it's a lot more enjoyable when you're with people you know and trust."*

(c) **Activity Planning and Designing:** Students emphasized the importance of well-organized activities that align with their interests and provide meaningful tasks. They expressed a preference for activities that allowed them to utilize their skills and talents, and that had a visible and tangible impact on the community.

- *"I really like when the activities are thoughtfully planned and align with our interests. It makes me feel like my time and effort are valued."*
- *"When I see the direct impact of what I'm doing, it's incredibly motivating."*
- *"It is great when the activities are meaningful, I know that my efforts makes a difference, I'm not just there for college work."*

(d) **Reflection Opportunities:** Students emphasized that by incorporating structured reflection activities, such as journals, group discussions, and presentations, students could process their experiences and recognize the personal and societal benefits of their community service.

- *"Taking the time to reflect on what we've done really helps put things into perspective. It's like we're learning and growing from our experiences, which makes it all feel more worthwhile."*
- *"Reflecting in our journals on our community service experiences helps me understand the broader significance of what we're doing. It's not just about completing tasks; it's about learning and growing as individuals while contributing to something greater than ourselves."*

(e) **Barriers to Participation:** Students univocally shared that in order to facilitate their opportunity to participate in community service, the college must address potential barriers to participation. This includes providing transportation, covering costs associated with service activities, and accommodating students' schedules.

- *"Transportation and scheduling the activities are real obstacles. If the college and community could work together with those issues, I think more students will like to participate and not shy away from these."*

IMPLICATIONS FOR EFFECTIVE IMPLEMENTATION

The findings highlight the complexity of students' attitudes towards community service and the factors that influence their participation. While the majority of students recognize the value of community service, their willingness to engage is significantly affected by how activities are structured and supported. Hence the following Implications for Effective Implementation have been suggested.

1. **Emphasize Intrinsic Benefits:** Activities should highlight the personal and societal benefits of community service, fostering a sense of intrinsic motivation among students.
2. **Balance Mandatory and Voluntary Elements:** While mandatory participation could ensure broad involvement, providing opportunities for voluntary engagement could enhance motivation and positive attitudes. Offering a mix of required and optional community service activities could cater to different motivations and preferences, ensuring that students feel a sense of choice and agency.
3. **Enhance Institutional Support:** Colleges should actively promote community service through encouragement from teachers and the provision of resources and logistical support. This includes recognizing and celebrating students' contributions, providing transportation and materials, and integrating community service into the broader educational curriculum.
4. **Design Engaging Activities:** Community service activities should be meaningful and aligned with students' interests to ensure a positive and enriching experience. Colleges should collaborate with students to identify their interests and passions, and design activities that allow students to apply their skills and make a tangible impact on their communities.
5. **Leverage Peer Influence:** Encouraging group participation could increase students' willingness to engage in community service through the influence of peer support. Colleges could facilitate group projects, create opportunities for peer mentorship, and promote a sense of community and camaraderie among students.
6. **Provide Reflection Opportunities:** Incorporating structured reflection activities, such as journals, group discussions, and presentations, could help students process their experiences and recognize the personal and societal benefits of their community service. Reflection allows students to connect their experiences to their personal values and goals, reinforcing the positive impact of their contributions.
7. **Tailoring Activities to Diverse Student Needs:** Recognizing that students come from diverse backgrounds with varying interests and capacities is critical for the success of community service activities. Activities should be flexible and adaptable, offering a range of activities that cater to different skills, interests, and availability. For instance, students interested in environmental issues might engage in community clean-ups or conservation projects, while those with a passion for social justice might work with local advocacy groups or support services for marginalized populations.

8. **Partnering with Community Organizations:** Colleges should establish strong partnerships with local community organizations to ensure that community service activities are well-coordinated and impactful. These partnerships could provide students with access to a wider array of service opportunities and ensure that their efforts are aligned with the needs of the community. Community organizations could also offer valuable resources and expertise, enhancing the educational value of the service experience.
9. **Encouraging Long-term Engagement:** Short-term community service projects could be beneficial, but encouraging long-term engagement could lead to deeper learning and more significant impacts. Colleges should provide opportunities for students to engage in ongoing projects or to return to the same service sites over multiple years. This continuity allows students to build stronger relationships with the community and to see the long-term effects of their contributions.
10. **Assessing and Recognizing Student Contributions:** Assessment and recognition are important components of effective community service activities. Colleges should implement mechanisms to assess the impact of students' service activities, both on the community and on the students themselves. This could include reflective essays, presentations, or project reports. Recognizing students' efforts through awards, certificates, or public acknowledgments could also reinforce the value of their contributions and motivate continued participation.
11. **Addressing Barriers to Participation:** To ensure that all students have the opportunity to participate in community service, Colleges must address potential barriers to participation. Colleges should also be mindful of the diverse needs of their student population, including students with disabilities, and ensure that community service opportunities are accessible to all.

CONCLUSION

Students' attitudes towards community service play a crucial role in the effectiveness of such activities. By understanding and addressing the factors that influence these attitudes, educators and policymakers could design and implement community service initiatives that not only fulfil educational goals but also foster a lifelong commitment to civic engagement. The insights gained from this study highlight the importance of intrinsic motivation, meaningful planning and designing of these activities, institutional support, and the role of peer influence in shaping positive attitudes towards community service.

By implementing the recommendations derived from this study, educational institutions could foster a more positive and effective community service culture among students, ultimately contributing to the development of socially responsible citizens. A collaborative approach, involving students, educators, community organizations, and policymakers, is essential for creating community service activities that are meaningful, engaging, and impactful. Through such efforts, community service could become a transformative experience that not only benefits the community but also enriches the lives of students, preparing them to become active and compassionate citizens in an increasingly interconnected world.

REFERENCES

Billig, S.H. (2000). Research on K-12 School-Based Service-Learning: The Evidence Builds. *PHI Delta Kappan*, 81(9), 658-664.

Kashyap, A. (2011). Community service and its impact on the personal and professional development of teacher trainees. *Indian Journal of Teacher Education*, 3(2), 45-57.

Metz, E., & Youniss, J. (2003). Longitudinal Gains in Civic Development through School-Based Required Service. *Political Psychology*, 24(3), 435-458.

Sankar, D. (2013). Perceptions of pre-service teachers towards community service-learning in India. *Educational Quest*, 4(3), 171-179.

Serow, R.C. (1991). Students and Voluntarism: Looking into the Motives of Community Service Participants. *American Educational Research Journal*, 28(3), 543-556.

Stukas, A. A., Snyder, M., & Clary, E. G. (1999). The Effects of "Mandatory Volunteerism" on Intentions to Volunteer. *Psychological Science*, 10(1), 59-64.

Thakur, A., & Pandey, R. (2015). Enhancing teacher education through community-based learning: An Indian perspective. *Journal of Community Service & Engagement in Higher Education*, 7(1), 35-50

Youniss, J., & Yates, M. (1997). Community Service and Social Responsibility in Youth. University of Chicago Press.

CHAPTER 13

Blending Traditional and Modern Pedagogies in Entrepreneurship Education: Cultivating an Entrepreneurial Mindset for Atmanirbhar Bharat

MANAV SHARMA

Teacher, Vibgyor High, Vadodara, Gujarat, India

ABSTRACT

This paper aims to explore the new blended approach of implementing conventional Indian teaching techniques along with modern methods to improve entrepreneurship learning in India. The goal is to develop an entrepreneurial outlook in students in association with the vision of Atmanirbhar Bharat and the objectives of NEP 2020 which focuses on the development of self-employment and skills. The old Indian education system like the Gurukul system which has principles like an integrated approach, individualized attention, and learning through experience and practice is compared with the new education system like project-based learning, use of technology, and collaborative learning environment. This paper also seeks to highlight the integration and advantages of incorporating these teaching methods by conducting a literature review and case studies. To identify key lessons and recommendations for practice and policy, the promising practices and effective strategies for integration in different educational settings are reviewed. The paper reveals that conventional approaches are capable of inculcating discipline, tenacity, and ethical values among students, while progressive methods offer creativity, problem-solving abilities, and practicality.

The difficulties of implementing these pedagogies are also reviewed, including barriers to change and resource constraints, with recommendations drawn from research findings and best practices. Based on the findings, it can be concluded that the integration of a traditional and online approach in the teaching and learning process can greatly improve the delivery of entrepreneurship education, equip the learners to thrive in a competitive economy, and help build a self-sustaining India. The following recommendations are made for educators and policymakers to effectively apply blended teaching methodologies that will promote capable future entrepreneurs under NEP 2020 entrepreneurship education.

Keywords: *Blended Approach, Entrepreneurship Education, Atmanirbhar Bharat, NEP 2020, Traditional and Modern Pedagogy*

INTRODUCTION

Introduction to NEP 2020

The National Education Policy (NEP) formulated in 2020 is a significant shift in the educational system in India as it envisions the country's learning system to meet the emerging needs of the 21st century. As a comprehensive and integrated framework for education, NEP 2020 aims to provide a framework for the enhancement of education across the country for the twenty-first-century learner that is inclusive of, inter alia, innovation, critical thinking, and problem-solving. Another major focus of NEP 2020 is the cultivation of an entrepreneurial disposition among students, consistent with the overall national liberalization agenda of Atmanirbhar Bharat, or self-reliant India. By emphasizing entrepreneurship education, the policy seeks to foster a culture of innovation and self-employment, that is, people being able to create their jobs.

NEP 2020 emphasizes the balanced use of indigenous knowledge and knowledge from the current world in the education system for the production of holistic human beings capable of excelling in the ever-evolving global economy. It replaces the conventional method of memorizing and reproducing information and instead focuses on learning through experience, problem-solving, and acquisition of skills, which in turn cultivates creativity, flexibility, and perseverance among learners. In line with the sub-theme of Entrepreneurship and Job Creation, NEP 2020 outlines guidelines through which educational institutions can adopt entrepreneurship education as an integral part of the curriculum to produce entrepreneurs who will foster economic development in India.

While implementing NEP 2020, educators need to understand how conventional and current pedagogical practices can be integrated within the context of entrepreneurship education. It has the potential to develop an entrepreneurial mindset along with building a generation of independent and employable citizens ready to carve their destiny for a better and self-reliant India.

Key Timelines of NEP 2020

The National Education Policy (NEP) 2020 has laid down a broad vision for the development of education in India and it is rightly stated that the successful implementation of the policy is the key to achieving the vision. In this regard, NEP 2020 has laid out a structure with timelines to ensure that the implementation of the policy's various recommendations happens in phases.

Key timelines for implementation in NEP 2020 include: Key timelines for implementation in NEP 2020 include:

1. **Short-term goals (0–2 years):** To do this, more emphasis should be placed on short-term strategies like developing new curricular frameworks, reviewing existing policies, and starting experimental projects for the application of new educational models. In this phase, activities focus on raising consciousness, mobilizing support to sustain policy changes, and establishing conditions for more extensive reforms.
2. **Medium-term goals (3–5 years):** This phase entails expansion of such pilot interventions, improvement of structures, and development of human resources among the instructors and other officials in schools. Curriculum reforms, including the incorporation of entrepreneurship education, are advanced and enacted across various learning institutions.
3. **Long-term goals (6–10 years and beyond):** The long-term goals of NEP 2020 lay in the complete realization of the change process that has been set into motion, and the creation of rich learning environments that foster creativity, critical thinking, and innovation. The policies may be adjusted regularly by the monitoring and evaluation results to ensure that they are relevant to the current needs and challenges.
4. **Continuous review and adaptation:** According to the NEP 2020, policies should be reviewed periodically and revised to reflect the current state of the nation and the goals of the country to ensure that they are relevant and effective in delivering the intended goals. It is common

to find feedback mechanisms used to elicit feedback from the stakeholders to enhance the quality and explore new ideas.

Therefore, by following these timelines and having a phased methodology of implementation of NEP 2020, it seeks to bring the desired change in the education system of India and turn, play a part in the fulfillment of the larger vision of Atmanirbhar Bharat by nurturing the young generation to be more enterprising and innovative to drive the economic and social progress of the country.

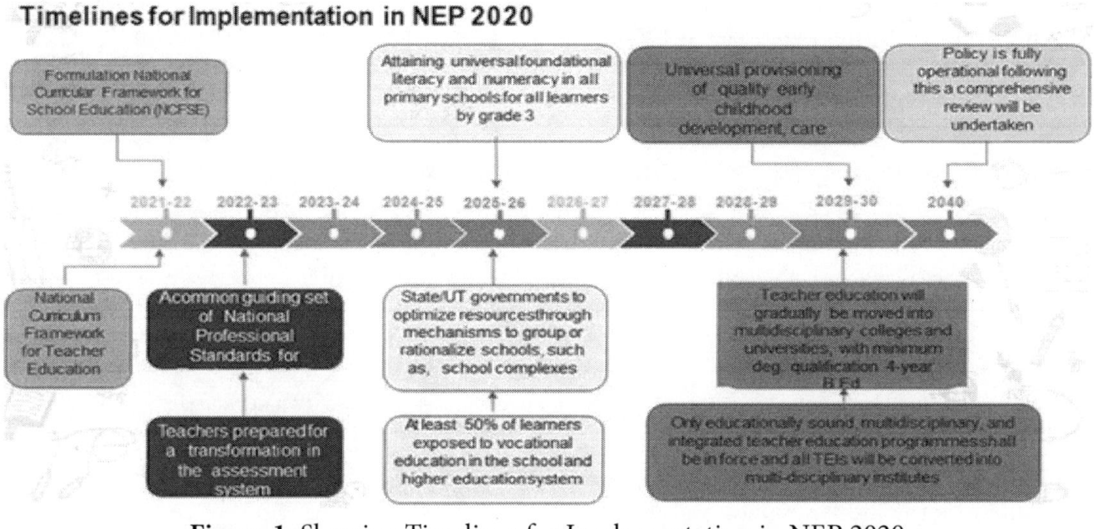

Figure 1: Showing Timelines for Implementation in NEP 2020

Traditional Indian Pedagogies

The Indian education system, especially the Gurukul system, can inspire the current generation of teachers and students with its rich set of methodologies which are still effective in the present-day world. The Gurukul system of education has been derived from the historical Indian culture and it aims at the integrated personality development of an individual. This system in which students also resided with their teacher or guru, was a seamless way of learning that incorporated the practicality of everyday life with the concepts taught in class.

Integrated Approach

This was one of the fundamental features of the Gurukul system of education as was mentioned earlier. In contrast to the modern approaches to education that are divided into subjects and classes, the Gurukul system of education did not restrict learning to the four walls of a classroom. However, the training was integrated with the daily tasks and procedures so that the theoretical aspect and the practical one were combined. Through engaging in different tasks, which ranged from farming, crafting, or performing administrative work, the students were able to see the real-life application of what they were learning in class. This method enabled students to have a holistic view of the environment and instilled in them the virtues of responsibility and independence.

The integrated approach whereby the students were allowed to participate in their learning process in an active manner rather than just receiving information made the learning process effective. They could then relate to what they were learning by observing how such concepts could be applied in their daily lives, and this made their learning more effective and lasting. This principle can be especially helpful in modern entrepreneurship education when the ability to apply theoretical knowledge in practice is of great importance for the building of entrepreneurial competencies.

Individualized Attention

The third characteristic of the Gurukul system was that the guru used to guide the students personally. Every learner benefited from a personal approach, which was given in accordance with his or her abilities and difficulties. This one-on-one mentoring made it possible to facilitate learning depending on the needs of the student and thus promoted their development.

This was made possible by the close and personal bond between the guru and the student. The guru, who could be a teacher, a guide, and even a nurturing parent, saw the essence and possibilities of each learner. This deep understanding allowed the guru to be more specific in his or her intervention and to assist students in addressing their problems while at the same time strengthening their assets.

It is thus important to note that individual attention is particularly useful in the process of entrepreneurship education. It is important to note that young people who want to start a business often have multiple strong skills, as well as weaknesses. This method is more effective in cultivating their entrepreneurial spirit by catering to each child's individual needs rather than following a standardized education system.

Experiential Learning

The Gurukul system was much more focused on the experiential learning process. The emphasis was put on the process of learning and not on memorization, and students were encouraged to practice on their own. This approach to learning also had a positive implication in the sense that it allowed the students to acquire practical skills and knowledge in their fields of study.

Gurukul's system of learning was a very effective one where learning was a process of active participation. They solved problems that demanded creativity, innovation, and critical thinking, all of which are valuable in entrepreneurship. Thus, by achieving the objectives and solving real-life problems, students gained the confidence and the competencies required to succeed in life and at work.

The concept of experiential learning is also relevant to meet the requirements of modern education for entrepreneurship. In particular, entrepreneurs should be able to use their knowledge in the real world, be creative, and make decisions on the spot.

Relevance to Modern Education

The idea of the Gurukul system, such as the integrated concept of learning, personal attention, and the emphasis on experience is not only relevant to the past but also to the present day and especially to the education of entrepreneurs. These principles create a more comprehensive and realistic learning environment that is needed to cultivate entrepreneurial abilities and attitudes.

The enhancement of these traditional approaches with modern teaching techniques and styles can help in the development of an effective and holistic entrepreneurship education curriculum. By incorporating the best aspects of the Gurukul system such as its focus on a child's personality or character along with modern approaches like project-based learning and technology integration, educators can create a program that will make students into future entrepreneurs who are innovative, adaptable, and self-sufficient.

In conclusion, the traditional Indian education system is beneficial for modern entrepreneurship education as it provides a number of important pedagogical ideas. By incorporating the best of both worlds, educators can help nurture the spirit of entrepreneurship among the learners in accordance with the Atmanirbhar Bharat and the NEP 2020 goals. This hybrid model can assist in the development of a new generation of entrepreneurs who will effectively support the Indian economy and its emergent self-sufficiency.

REVIEW OF LITERATURE

Nandram, S. S., Joshi, A., Sukhada, N., & Dhital, V. (2021): The research paper "Delivering holistic education for contemporary times: Banasthali Vidyapith and the Gurukula system" offers valuable insights into its potential for cultivating well-rounded individuals equipped with entrepreneurial skills for contemporary challenges. By integrating traditional principles with modern pedagogies, Banasthali Vidyapith exemplifies a holistic approach to education, fostering creativity, critical thinking, and ethical values essential for Atmanirbhar Bharat. While the institution faces challenges of scalability and alignment with industry needs, ongoing research endeavors seek to address these concerns, emphasizing the importance of collaborative efforts in realizing the full potential of such educational models.

Soni, B.K., & Trivedi, J.C. (2018): The paper tells about an intriguing case study on Gotirth Vidyapeeth, which draws inspiration from the traditional Indian Gurukul system rooted in the Vedas. Emphasizing oral transmission of knowledge overwriting, the Gurukul setup aims to develop cognition, intelligence, and memory in students. Founded by Shri Gopalbhai Sutaria in Ahmedabad, Gotirth Vidyapeeth offers a holistic education in a spiritual environment, reminiscent of ancient traditions. Students engage in a wide array of practical skills, including music, martial arts, agriculture, and ayurvedic medicine preparation, alongside traditional subjects like recitation of shlokas and Vedic mathematics. Notably, the school fosters inclusivity by welcoming students from diverse socioeconomic backgrounds. This case study offers valuable insights into the revival of traditional educational methods and their relevance in contemporary society, particularly in promoting holistic development and preserving cultural heritage. Further exploration of the outcomes and challenges of such initiatives could provide valuable contributions to educational discourse.

Jagathkar, A. (2019): The research paper provides a thought-provoking exploration of human evolution alongside India's rich educational heritage. Drawing parallels between the gradual evolution of humanity from ape-like ancestors and the historical progression of education in India, the abstract highlights the enduring significance of oral transmission of knowledge and the revered 'Guru Shishya Parampara'. By elucidating how education was primarily imparted orally by sages and scholars, fostering deep bonds between teachers and students, the abstract underscores the profound legacy of traditional Indian education systems. It evokes images of schools nestled within hermitages, cottages, temples, and community centers, serving as vibrant hubs of learning and cultural exchange. This juxtaposition of human evolution and educational legacy prompts reflection on the enduring value of oral traditions and intimate mentorship in shaping societies across time. Further exploration of the implications of these historical insights for contemporary educational practices could offer valuable perspectives on nurturing meaningful learning experiences and fostering enduring teacher-student relationships in modern contexts.

Sharma, A., & Joshi, A. (2021): The research paper introduces an insightful exploration of sustainable management education through the lens of Bhartiya Gurukul's philosophy, shedding light on the indigenous wisdom that can enrich contemporary educational practices. By elucidating how the principles of Gurukul philosophy can foster sustainability in management education, the paper offers a nuanced understanding of five key facilitating factors: Aparigrah (minimalism), Swawlamban (self-reliance), Udyam (entrepreneurial initiatives), Uttardayitava (responsibility), and Chetna (consciousness). These factors not only resonate with traditional Indian values but also hold relevance for addressing contemporary challenges in management education. The paper's emphasis on implications for institutional educational philosophy underscores the transformative potential of integrating indigenous wisdom into educational frameworks. Furthermore, by delineating implications for internal and external stakeholders, including faculty, students, infrastructure, board members, corporate entities, and society at large, the paper offers actionable insights for fostering holistic and sustainable management education practices. Overall, this paper contributes to the discourse on educational sustainability by bridging traditional wisdom with modern management principles, paving the way for meaningful and impactful educational reforms in the context of evolving global challenges.

Joshi, A., & Gupta, R.K. (2017): The research paper offers a critical examination of the challenges facing education in Bharat, particularly concerning issues of quality and the legacy of colonial influence on traditional educational institutions like Gurukuls. By contextualizing the post-colonial educational landscape, the paper highlights the failure of post-independence efforts to decolonize education or revitalize traditional systems effectively. Against this backdrop, the paper adopts a post-colonial perspective to explore the contemporary relevance of Gurukuls through an ethnographic account of one such institution. Through this lens, the paper proposes actionable propositions for enhancing the quality of elementary education, drawing insights from the Gurukul model. This approach not only sheds light on the enduring presence of Gurukuls but also offers a pathway for reclaiming indigenous educational practices to address current educational challenges. The ethnographic account provides valuable empirical grounding for the proposed recommendations, enhancing their relevance and applicability. Overall, this paper contributes to the discourse on educational reform by advocating for a reevaluation of traditional educational systems within the context of post-colonial educational paradigms, offering potential avenues for improving educational quality in Bharat.

KEY FINDINGS

- **Comparative Analysis of Educational Systems:** The study also aims at comparing the conventional ways of teaching that are prevalent in India, especially the Gurukul system with the contemporary ways of teaching in the field of entrepreneurship. The analysis of this comparison highlights the advantages and limitations of each approach, which underlines the need to integrate them to improve the learning of entrepreneurship in India. In this paper, I have advocated for a balanced approach to teaching and learning, which incorporates both conventional and innovative approaches to education, as a way of equipping students for the challenges of entrepreneurship within Atmanirbhar Bharat.
- **Incorporating Indigenous Wisdom:** The study adopts a post-colonial perspective and examines how indigenous paradigms of learning and knowledge, as exemplified by the Gurukul system, can be incorporated into modern entrepreneurship education. This emphasizes the need to blend conventional wisdom with contemporary technologies in the light of Atmanirbhar Bharat and NEP 2020 vision. Thus, by engaging with indigenous knowledge, the paper argues for incorporating culturally appropriate and diverse learning approaches into the entrepreneurship curriculum to promote an entrepreneurial disposition imbued with Indian values.
- **Key Facilitating Factors:** Through a comprehensive literature review and analysis of case studies, the research identifies five facilitating factors for making management education more meaningful and sustainable: The main themes are minimalism, self-reliance, entrepreneurial initiatives, responsibility, and consciousness. These factors link conventional culture to contemporary business management theories and practices to provide a comprehensive model of entrepreneurship learning. In this manner, the paper posits that the noted facilitating factors can be nurtured by educators in order to foster the development of key competencies and orientations among learners, thereby enabling them to become successful entrepreneurs in a constantly changing economic environment.
- **Challenges and Recommendations:** The study focuses on the difficulties of introducing the integration of traditional and innovative approaches to teaching entrepreneurship such as resistance to change and lack of resources. However, it also presents recommendations for practice that can be useful for educators and policymakers to implement these methodologies. These recommendations are intended to help create capable future entrepreneurs under entrepreneurship education under NEP 2020. Therefore, based on the identified challenges, my paper recommends that stakeholders should adopt the following strategies in order to improve the quality and sustainability of entrepreneurship education in India.
- **Empirical Evidence and Practical Implications:** In the current study, based on the analysis of case studies and literature review, I contribute to the understanding of the outcomes of blended teaching approaches in the context of entrepreneurship education. It focuses on the application of the combined use of both traditional and modern teaching methods, as well as stressing the importance of such approaches in the development of entrepreneurship skills and the attainment of the goal of a self-reliant India. In line with the above rationale, this paper seeks to provide an empirical and practical perspective on the need to develop and

introduce new forms of teaching and learning entrepreneurship that will enable students to succeed in the new economy.

CONCLUSION

This research paper calls for an integration of both conventional and innovative teaching practices in the teaching of entrepreneurship for Atmanirbhar Bharat. By comparing the educational systems of the past and present, one can identify the benefits and drawbacks of both the traditional Indian teaching methodology, the Gurukul system, and the contemporary approach. Hence, the application of these approaches is essential for improving entrepreneurship learning in India. Further, the idea of integrating an indigenous knowledge system, as described in the Gurukul model, is imperative for placing entrepreneurship education in tune with the vision of Atmanirbhar Bharat and the spirit of NEP 2020. This way, by engaging with indigenous knowledge, educators can add culturally sensitive and diverse teaching-learning processes into the curriculum to develop an entrepreneurial mindset based on Indian ethos. The definition of the main factors of facilitation, such as minimalism, self-sufficiency, entrepreneurship, responsibility, and awareness, reveals the link between traditional culture and modern theories of business management. These are areas that educators can foster to ensure that the learners grow with the right competencies and orientation as entrepreneurs in the prevailing economic environment.

In light of the challenges of blended teaching methodologies implementation, such as resistance to change and resource constraints, this paper offers practical solutions for educators and policymakers. The following recommendations are meant to ensure capable future entrepreneurship education under NEP 2020 entrepreneurship education. Moreover, taking into consideration the findings of the empirical studies and the practical experience, it is possible to state that the combination of innovative and conventional approaches effectively contributes to the development of students' entrepreneurial competencies. Blended teaching methodologies prepare the student to face the competitive economy and help in building a self-sustaining India as it helps inculcate creativity, problem-solving ability, practicality, discipline, tenacity, and ethical values among the learners. In conclusion, this research emphasizes the need to adopt a middle approach in teaching and learning entrepreneurship, with a blend of traditional and innovative methods. By doing so, stakeholders can effectively prepare students to navigate the complexities of entrepreneurship and contribute to the realization of Atmanirbhar Bharat's vision.

REFERENCES

Codrops. (n.d.). *Philosophy of Indian Pedagogy: Philosophy of Indian Pedagogy - page: 98*. kireetjoshiarchives.com. https://www.kireetjoshiarchives.com/philosophy/philosophy-of-indian-pedagogy/philosophy-of-indian-pedagogy/&page=98

Ias, S. (n.d.). *National Education Policy 2020 : A Vision ahead*. SANSKRITI IAS. https://www.sanskritiias.com/current-affairs/national-education-policy-2020.

Jagathkar, A. (2019). Building India: A Country of Innovations. *IARS International Research Journal, 9*(1). https://doi.org/10.51611/iars.irj.v9i1.2019.99.

Joshi, A., & Gupta, R. K. (2017). Elementary education in Bharat (that is India): insights from a postcolonial ethnographic study of a Gurukul. *International Journal of Indian Culture and Business Management*, *15*(1), 100. https://doi.org/10.1504/ijicbm.2017.085390.

Nandram, S. S., Joshi, A., Sukhada, N., & Dhital, V. (2021). Delivering holistic education for contemporary times: Banasthali Vidyapith and the Gurukula system. *International Journal of Business and Globalisation*, *29*(2), 222. https://doi.org/10.1504/ijbg.2021.118235

Nic, L. P. (n.d.-a). *About National Education policy | Government of India, Ministry of Education*. https://www.education.gov.in/nep/about-nep#:~:text=National%20Education%20Policy%2C%202020%20(NEP,quality%20education%20to%20all%2C%20thereby

Nic, L. P. (n.d.-b). *Consultation process | Government of India, Ministry of Education*. https://www.education.gov.in/nep/tp-nep.

Sharma, A., & Joshi, A. (2021). Sustainable management education: insights from indigenous wisdom of Bhartiya (Indian) Gurukul philosophy. *World Review of Entrepreneurship, Management and Sustainable Development*, *17*(4), 471. https://doi.org/10.1504/wremsd.2021.116660

Soni, B.K., & Trivedi, J.C. (2018). A case study on Gurukul System of education. *Rajagiri Management Journal*, *12*(2), 39–63. http://journals.rajagiri.edu/index.php/rmj/article/download/283/259

CHAPTER 14

Real-Time Sign Language Interpretation Using Advanced Object Detection Algorithms

Anant Dev Kapoor

Scholar, Faculty of Engineering and Technology, SRM Institute of Science and Technology, Chennai, India

Abstract

This paper presents a real-time sign language recognition system that incorporates object detection using convolutional neural networks and You Only Look Once. The objective of this study is to develop a sign language recognition system that can accurately detect and interpret American sign language gestures in real-time. This study evaluates the performance of YOLOv5 and YOLOv8 models on an American Sign Language (ASL) dataset, focusing on their ability to recognize hand gestures. The dataset is divided into 72 images for testing, 1512 for training, and 144 for validation. Both models were trained for 80 epochs on Google Collab using a Tesla T4 GPU. YOLOv5 showed faster and more consistent loss reduction, achieving 95% precision, 97% recall, and 96% map @0.5. In comparison, YOLOv8 demonstrated higher object detection accuracy but required fewer iterations to reach minimum loss, indicating its potential for quick adaptation. Despite this, YOLOv8's confusion matrix displayed more discrepancies, suggesting greater susceptibility to unseen images and real-life applications. Practical testing revealed that while both models perform well, YOLOv8 provided more accurate predictions for certain gestures not properly identified by YOLOv5. However, YOLOv5 performed better with intermediate-quality videos, highlighting its robustness. Overall, the study suggests that YOLOv5 is more stable and efficient in training, while YOLOv8 excels in detection speed and accuracy, making the choice between them dependent on specific application needs and constraints.

Keywords: *YOLOv5, CNN, American Sign Language (ASL), Hand Gesture Recognition, Deep Learning Models*

INTRODUCTION

Sign language is an important mode of communication for individuals with hearing impairments. It allows them to express themselves, communicate with others, and participate fully in society. Over the years, technology has played a crucial role in the development of sign language recognition systems. Early systems relied on hand-crafted features and traditional methods, but recent advancements in deep learning have significantly improved the accuracy and effectiveness of these systems. Convolutional neural networks and You Only Look Once are two popular techniques used for object detection and pattern recognition in sign language recognition systems. However, there is still a need for further research and development to enhance the real-time capabilities and accuracy of these systems. Sign language serves as a vital mode of communication for individuals with hearing impairments, providing them with a unique and effective way to express their thoughts, emotions, and ideas. It allows them to communicate with others and participate fully in various aspects of society, including education, employment, and social interactions. Over the years, technology has played a crucial role in the development and advancement of sign language recognition systems. Early systems relied heavily on hand-crafted features and traditional methods, which often required manual intervention and were not always accurate or efficient. However, recent advancements in deep learning have significantly improved the accuracy, effectiveness, and overall performance of these systems. Techniques such as convolutional neural networks and You Only Look Once are now widely used for object detection and pattern recognition in sign language recognition systems. These advanced methods have contributed to more reliable and precise recognition of sign language gestures. Despite these advancements, there is still a need for further research and development to enhance the real-time capabilities and accuracy of these systems. Continued efforts in this area will ensure that sign language recognition systems become even more robust and efficient, providing better support for individuals with hearing impairments and facilitating more seamless communication in various contexts.

BACKGROUND AND RELATED WORK

Sign language stands out for its reliance on physical gestures rather than specific sounds, making it unique in communication methods. It is particularly notable for how it uses hand movements, facial expressions, and body language to convey meaning, distinguishing it from spoken languages that depend on auditory signals. Significant research has been dedicated to facial and gesture recognition in this domain, focusing on various techniques and technologies to improve the accuracy and usability of sign language recognition systems. Current methods in this field can be categorized into two main approaches, each with its own set of advantages and challenges. The first approach involves using external devices for sign recognition. These devices can include specialized gloves, motion sensors, or other hardware that directly captures the user's hand movements and translates them into digital signals. This method can be very accurate but often involves higher costs and less portability due to the need for additional equipment. On the other hand, the second approach utilizes deep learning algorithms for detecting and recognizing hand gestures. This method relies on software and camera systems, making it more cost-effective and portable, but it can be more challenging to achieve high accuracy and low latency.

Various researchers have introduced innovative techniques for sign language (SL) recognition, addressing factors like cost, latency, performance, and portability. For instance, A. Bhattacharya discusses training a classifier on a dataset comprising 24 gestures that can be easily represented using the "bag of visual words" approach. This method involves identifying features in the images, clustering them to form a dictionary, and then representing each image as a histogram indicating the frequency of observed gestures. This approach allows for efficient recognition and categorization of gestures based on visual data.

T. Starner presents two real-time hidden Markov model-based systems that require only a camera to track the user's movements and recognize intermediate-level continuous American Sign Language (ASL). The original system, equipped with a desk-mounted camera, achieved a 92% accuracy rate. In contrast, the second system, which involved mounting the camera on the user's hat, reached an impressive 98% accuracy. These systems demonstrate the potential of using simple camera setups to achieve high accuracy in real-time sign language recognition. L. Aziz provides an extensive overview of recent advancements in visual object detection using deep learning. This review encompasses approximately 300 methods, including both region-based and classification-based object detection methods. Additionally, the authors explore publicly available benchmark datasets, discussing their origin, usage, advantages, limitations, and evaluation metrics. This comprehensive review highlights the significant progress made in the field and the importance of benchmark datasets for evaluating and improving recognition systems.

Furthermore, D. Naglot applies a Multi-layer Perceptron neural network to recognize several signs using a dataset comprising 520 samples. This method involves training a neural network with multiple layers to process and learn from the input data, enabling it to accurately recognize and classify various sign language gestures. The use of a Multi-layer Perceptron demonstrates the effectiveness of deep learning techniques in handling complex recognition tasks. In summary, the field of sign language recognition has seen substantial advancements through the use of both external devices and deep learning techniques. While each approach has its own set of challenges, ongoing research continues to improve the accuracy, efficiency, and portability of these systems. By leveraging advanced algorithms, innovative hardware solutions, and comprehensive datasets, researchers are making significant strides in creating robust and reliable sign language recognition systems that can facilitate better communication for individuals with hearing impairments.

METHODOLOGY

The primary datasets commonly utilized for object detection and segmentation are Pascal VOC 2007 and Microsoft COCO. These datasets have become standard benchmarks in the field due to their comprehensive annotations and variety of object categories. In this study, our focus is on YOLO versions 5 and 8, known for their superior performance in recognizing sign language

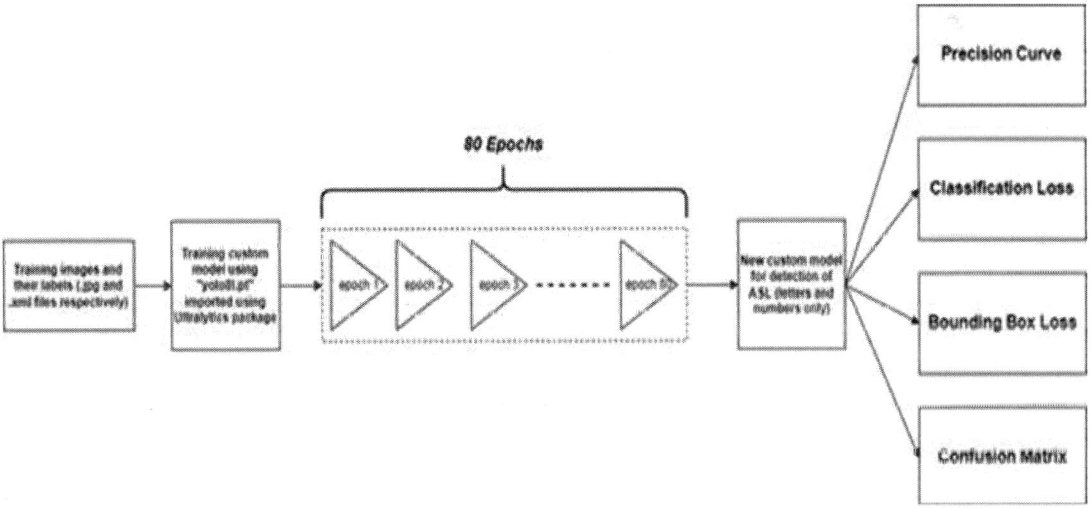

Figure 1: Architecture of the Proposed System Combining CNN and YOLOv5 for Real-Time Sign Language Recognition with Object Detection

with high accuracy. The methodology proposed is illustrated in Figure 1, showcasing the steps and processes involved in our experimental setup and execution.

YOLOv5, a lightweight and fast deep learning architecture, serves as the foundation for our experiments. Developed by Glenn Jocher, YOLOv5 boasts reduced computational requirements compared to other state-of-the-art models while maintaining high accuracy. Unlike previous versions, YOLOv5 was introduced without a supporting paper, indicating ongoing development and community-driven improvements. The architecture, released in June 2020, is constructed using Python, which simplifies installation and integration, and leverages the expansive PyTorch community for support and future development. YOLOv5 employs CSPNET as its backbone for feature extraction and integrates Path Aggregation Network (PANet) to enhance information flow, thereby improving detection performance.

We opt for YOLOv5 for several reasons:

1. YOLOv5 incorporates state-of-the-art features such as advanced activation functions, well-tuned hyperparameters, effective data augmentation techniques, and user-friendly manuals. These features collectively contribute to its high performance and ease of use.
2. Its simple architecture facilitates computationally efficient training, even with limited resources, making it accessible for researchers and developers with varying levels of computational power.
3. YOLOv5's compact and lightweight design makes it well-suited for mobile devices and embedded applications, allowing for real-time object detection in resource-constrained environments.

YOLOv8, also developed by Glenn Jocher and introduced on January 23, 2023, is the latest addition to the YOLO family. This version is still undergoing development, but it introduces new and innovative features like Anchor-free detection, which simplifies the detection process by removing the need for predefined anchor boxes, and mosaic augmentation, which enhances training data diversity by combining multiple images into one during the augmentation process. YOLOv8 comes with a Command Line Interface (CLI) for easier model training comprehension and a Python package for smoother development experiences compared to its predecessors. This combination of features and tools makes YOLOv8 a powerful option for cutting-edge object detection tasks.

Regarding the loss function, it is calculated by summing the results of various bounding box parameters' loss functions. The equations involve terms representing coordinates, dimensions, and confidence scores, with detailed explanations provided for each component. This comprehensive approach ensures that the model is accurately penalized for errors in object localization and classification, leading to improved performance over time. The installation and initialization process for both YOLOv5 and YOLOv8 are outlined, with a focus on running simultaneous experiments on different devices for efficient comparison. By doing so, we can directly compare the performance and efficiency of each version under similar conditions, providing valuable insights into their relative strengths and weaknesses. For our experimentation, we select the "American Sign Language letters" dataset from publicly available datasets on Roboflow. This dataset is well-suited for our purposes as it contains a diverse set of images representing the various letters of the American Sign Language alphabet. PyTorch and Ultralytics packages are utilized to implement the YOLO algorithms effectively. The process of downloading and implementing the dataset into the model is described in detail, highlighting the integration of the Roboflow package for easy dataset management and the importance of the Ultralytics package for ensuring compatibility with YOLO algorithms.

In summary, our study leverages the strengths of YOLOv5 and YOLOv8 to advance the field of sign language recognition. By using robust datasets and state-of-the-art techniques, we aim to achieve high accuracy and efficiency in recognizing sign language gestures. The combination of advanced deep learning architectures, efficient data handling, and thorough experimentation provides a solid foundation for future research and development in this important area.

RESULTS

The dataset utilized for training, testing, and validation comprises images divided in a ratio of 1:21:2, equating to 72 images for testing, 1512 images for training, and 144 images for validation. This specific distribution ensures that the model has ample data for learning while retaining enough for evaluation and fine-tuning. The new model underwent an extensive training process for 80 epochs, ensuring thorough learning of the patterns and nuances within the dataset. In Figure 2, the confusion matrix for YOLOv5 and YOLOv8 on the American Sign Language (ASL) dataset is depicted. Notably, compared to YOLOv5, YOLOv8's confusion matrix reveals more discrepancies, rendering YOLOv8 more susceptible to unseen images and real-life applications. In YOLOv5, there are 12 occurrences where the predicted values differ from the true values, indicating a relatively high accuracy in predictions. However, in YOLOv8, these occurrences increase slightly to 13, suggesting that while YOLOv8 is highly capable, it may still encounter more challenges with unseen data or in practical scenarios. A significant difference between YOLOv5 and YOLOv8 is the number of iterations required to reach the minimum loss; YOLOv8 achieves this in fewer iterations than YOLOv5, possibly due to YOLOv8 being in its early developmental stages and benefiting from more recent advancements and optimizations. Additionally, YOLOv8

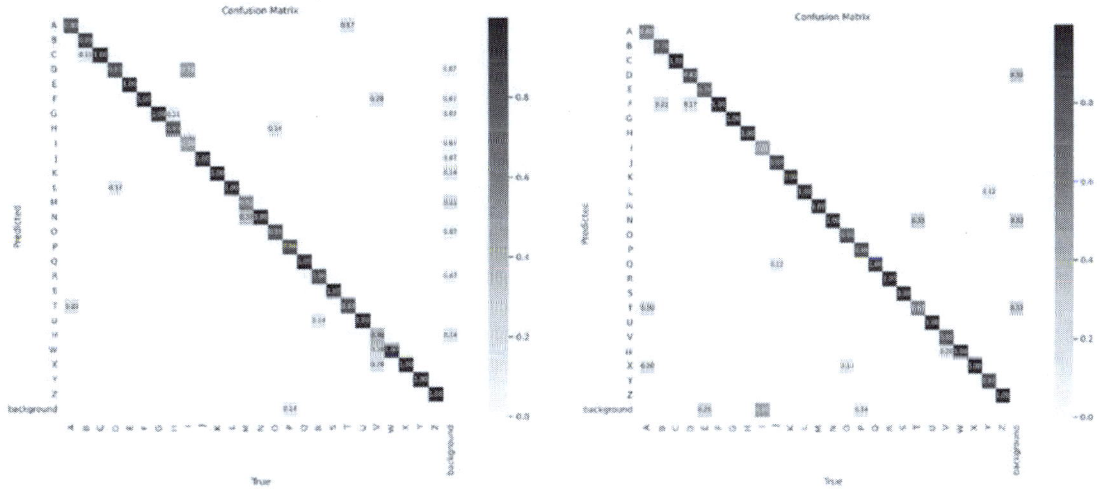

Figure 2: Confusion Matrix for YOLOv5 and YOLOv8

exhibits greater significance in specific wrong predictions, with the model presenting seven such cases where the errors are more pronounced compared to YOLOv5.

Figure 3 illustrates the comparison of loss reduction between both YOLO versions. YOLOv5 demonstrates a faster and more consistent descent in both bounding box and classification loss compared to YOLOv8. This indicates that YOLOv5 may be more stable and efficient during the training process. The custom model attained 95% precision, 97% recall, and 96% mAP @0.5, showcasing its capabilities in real-time hand gesture recognition. These metrics highlight the model's high accuracy and reliability in detecting and interpreting sign language gestures. When comparing the bounding box and classification loss, YOLOv8 outperforms YOLOv5 in object detection tasks, demonstrating a higher accuracy in identifying and localizing objects within the images. Conversely, YOLOv5 excels in classification accuracy, meaning it is more precise in correctly identifying the class of the detected objects. Overall, YOLOv8 proves to be more accurate and faster in object detection and hand gesture recognition, making it a suitable choice for applications requiring quick and reliable detection.

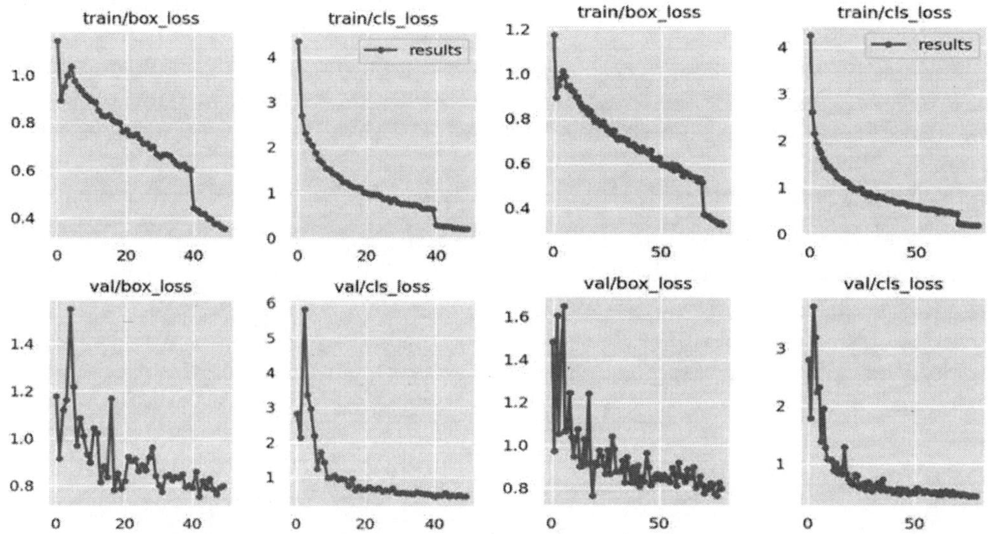

Figure 3: Loss Reduction for YOLOv5 (left) and YOLOv8 (right)

After the training phase, a small batch from the validation subset was tested, revealing that both models perform well with distinct gestures. However, YOLOv8 provides more accurate predictions, including certain gestures that are not properly predicted by YOLOv5. This suggests that YOLOv8 may have a better generalization capability and can handle a wider variety of gestures more effectively.

Table 2 presents the initial and final loss values for bounding box loss, classification loss, and mean Average Precision (mAP) for both YOLOv5 and YOLOv8. Both neural networks were trained for 80 epochs on Google Colab with a Tesla T4 GPU at precision level 8 (P(8)) and using CUDA version 12.0. The training process took approximately 6 hours, excluding the time needed for dataset preparation and testing. This efficient training duration highlights the capabilities of modern GPUs and optimized training pipelines.

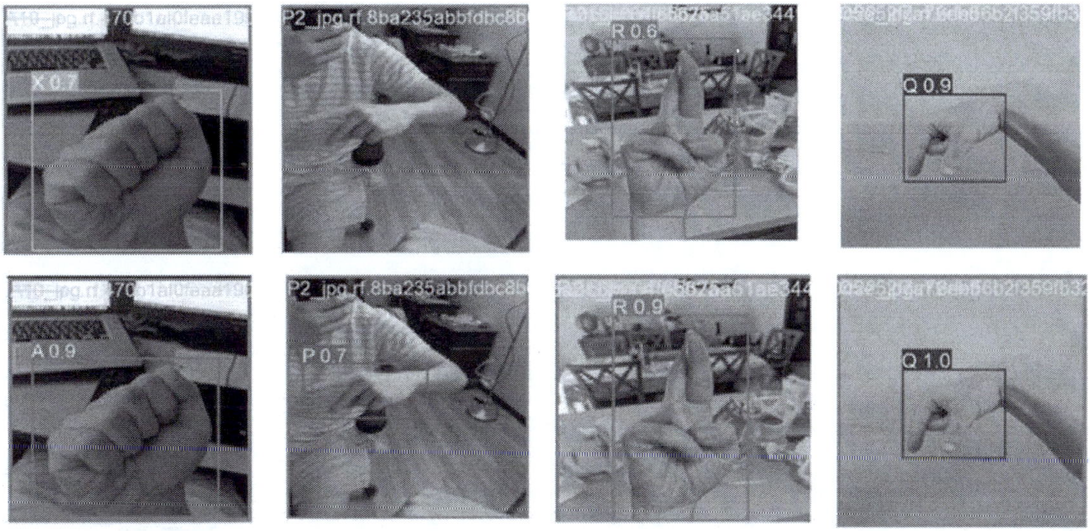

(a) Validation Result (YOLOv5) (b) Validation Result (YOLOv8)

Figure 4: Validation Result (YOLOv5) and Validation Result (YOLOv8)

Figures 4a, 4b, 5a, and 5b illustrate the validation results and precision curves for both models. While the performance of both models is excellent with minimal differences, practical testing reveals that YOLOv5 performs slightly better, particularly when testing with intermediate-quality recorded videos. This suggests that YOLOv5 may be more robust in handling variations in video quality, making it a reliable choice for real-world applications where video quality can vary. In summary, the comparative analysis of YOLOv5 and YOLOv8 using the ASL dataset demonstrates the strengths and weaknesses of each model.

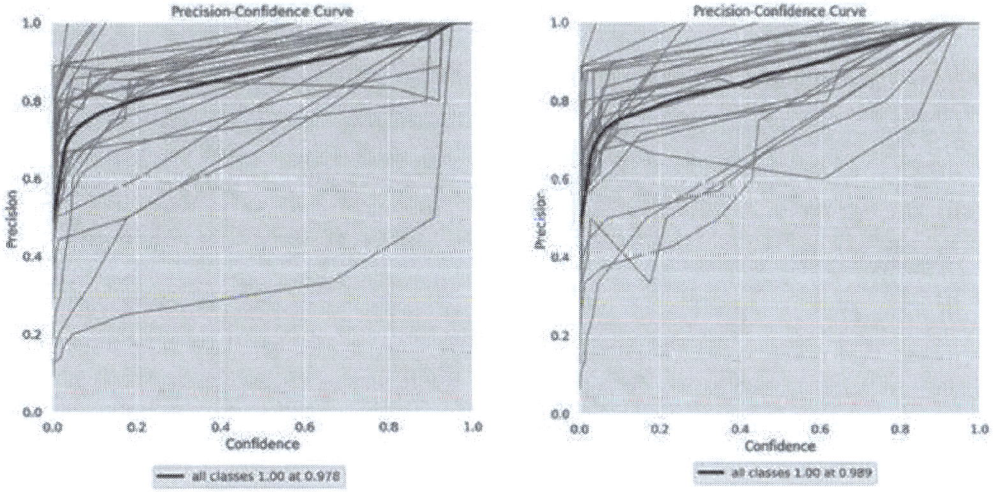

(a) Precision Curve (YOLOv5) (b) Precision Curve (YOLOv8)

Figure 5: Precision Curve (YOLOv5) and Precision Curve (YOLOv8)

CONCLUSION

While YOLOv8 exhibits superior performance during training and validation to a certain degree, it also shows promising results in testing. Nevertheless, the model is not yet suitable for practical applications involving constant frame movement, positioning YOLOv8 as more of a conceptual model in its current state. Conversely, YOLOv5 adeptly predicts the correct gesture and even fills in predictions left blank by the other model.

REFERENCES

Aziz, M. S. B. H. S., Sheikh, U. U., & Ayub, S. (2020). Exploring deep learning-based architecture, strategies, applications, and current trends in generic object detection: A comprehensive review. *IEEE Access*, 8, 170461-170495.

Bhattacharya, A., Zope, V., Kumbhar, K., Borwankar, P., & Mendes, A. (2019). Classification of sign language gestures using machine learning. *Image* 8, no. 12.

Chuan, C.-H., Regina, E., & Guardino, C. (2014). American Sign Language Recognition Using Leap Motion Sensor. *2014 13th International Conference on Machine Learning and Applications.*

Daniels, S., Suciati, N., & Fathichah, C. (2021, February). Indonesian sign language recognition using YOLO method. In *IOP Conference Series: Materials Science and Engineering* (Vol. 1077, No. 1, p. 012029). IOP Publishing.

Halvardsson, G., Peterson, J., Soto-Valero, C., & Baudry, B. (2021). Interpretation of Swedish sign language using convolutional neural networks and transfer learning. *SN Computer Science*, 2(3), 207.

Imagawa, S. L., & Igi, S. (1998). Color-based hands tracking system for sign language recognition. Proceedings Third IEEE International Conference on Automatic Face and Gesture Recognition.

Naglot, D., & Kulkarni, M. (2016). Real-time sign language recognition using the leap motion controller. *2016 International Conference on Inventive Computation Technologies* (*ICICT*).

Starner, J. W., & Pentland, A. (1998). Real-time American sign language recognition using desk and wearable computer-based video. *IEEE Transactions on Pattern Analysis and Machine Intelligence*, 20(12), 1371-1375

Vedak, O. et al. (2019). Sign language interpreter using image processing and machine learning. *International Research Journal of Engineering and Technology* (*IRJET*).

World Health Organization (2023, February 27). Deafness and hearing loss. Retrieved from: https://www.who.int/news-room/fact-sheets/detail/deafness-and-hearing-loss.

CHAPTER 15

Financial Literacy and Inclusion in India with Reference to Digital Payment

DR. YOGITA DWIVEDI

Assistant Professor, JBS, Jaypee Institute of Information Technology University, Noida, India

ABSTRACT

The digital payment revolution has significantly transformed the way individuals and businesses conduct financial transactions. Customer awareness and adoption of digital payment methods have played a crucial role in driving this revolution. In times of crisis, like the COVID-19 pandemic, when physical contact is restricted, stress the value of digital payments. Financial literacy plays a crucial role in people's ability to access financial services through digital payment. Indian financial literacy has been examined in several studies. Financial inclusion is one of the most crucial elements in the fight against poverty. Achieving financial literacy is all about making informed decisions about spending and managing money while making the necessary educated assumptions. The study investigates the variables that affect financial literacy in India as well as the relationship between financial literacy and saving and investing habits. With the objective, a wide range of topics is discussed, including the definition, history, and applications of financial literacy in India. The study concluded that to eradicate poverty and foster prosperity, a comprehensive, long-term education program should be widely available to rural communities. The study suggests that there is a need for targeted education and awareness programs, and tailored campaigns can address the knowledge gap and encourage adoption.

Keywords: *Digital Payment, Financial Literacy, Financial Inclusion, Education Programme, Rural Communities*

INTRODUCTION

The growth of digital payments in India has been remarkable in recent years, driven by various factors such as government initiatives, technological advancements, increasing internet and smartphone penetration, and changing consumer behaviour. The volume of digital payment transactions in India has witnessed significant growth across various channels, including UPI, mobile wallets, debit/credit cards, and internet banking. According to data from the Reserve Bank of India (RBI), the total volume of digital transactions in India has been steadily increasing year-over-year. Financial literacy levels in India have been improving gradually, but there are still significant gaps in awareness and understanding of financial products and services, particularly among rural and underserved populations. Various initiatives by the government, financial institutions, and non-profit organizations aim to promote financial literacy and education, including awareness campaigns, training programs, and digital literacy initiatives. Increasing financial literacy is essential for promoting financial inclusion, empowering individuals to make informed decisions, and fostering overall economic development. Financial literacy and financial inclusion are crucial components of economic development, especially in a country as diverse and dynamic as India.

LITERATURE REVIEW

Seldal & Nyhus (2022), explained that COVID-19 has caused an increase in online shopping that has made digital payments more essential. Men, highly educated individuals, high earners, and younger generations are more likely to use new payment technologies. Additionally, non-users of new payment technologies were less financially literate and more financially vulnerable. In regard to financial literacy, gender, and financial vulnerability, they have rejected their findings. According to Khan et al. (2022), financial literacy has a big impact on financial inclusion since it makes financial products easier for people to understand and increases demand for them. Financial inclusion is influenced by financial education initiatives, affordability incentives, and sociodemographic characteristics like age, education level, and attitude. In addition to be a moderator, financial literacy also affects institutional and cognitive aspects. Azeez & Akhter (2021), concluded that among rural areas, determinants such as income, employment, gender, land ownership, and ration cards have a significant impact on digital financial literacy. Along with the efforts of financial institutions, governments should frame policies that will increase income and provide better employment opportunities.

McKillop, French, and Stewart (2020), assess whether smartphone apps can be utilised to improve desirable financial capability. In a randomised control trial, the authors find significant improvements in 'financial knowledge, understanding and basic skills' and 'attitudes and motivations' for the group of individuals that used the apps. Those using the apps were more likely to keep track of their income and expenditure and proved to be more resilient when faced with a financial shock. The ability of business owners to secure the funding they require has a favorable impact on the expansion of the national economy (Kim, et al., 2018). In order to safeguard customers, promote financial stability and capacity, and encourage financial inclusion, financial literacy is essential (Iriobe et al., 2017). Making sure that vulnerable populations, like those with low incomes, have access to suitable financial products and services at a reasonable price is known as financial inclusion (Mohamad & Nasir, 2019; Deka, 2015).

RESEARCH METHODOLOGY

The study is descriptive in nature. Secondary data has been used to understand and analyse the relationship between financial literacy and financial inclusion in India. The study investigates the variables that affect financial literacy in India as well as the relationship between financial literacy and financial adoption.

OBJECTIVE

The study based on following objectives:

1. To understand the financial literacy and adoption of digital payment.
2. To investigate the digital payment growth in India.
3. To analyse the future of digital payment in India 2027.

FINANCIAL LITERACY AND FINANCIAL INCLUSION

Financial literacy refers to understanding financial concepts like budgeting, saving, investing, borrowing, and debt management. It involves making wise financial decisions and navigating the financial services and goods. Financial inclusion aims to provide access to appropriate and reasonably priced financial products and services, including credit, insurance, and banking services, to all societal segments, especially those historically underserved or excluded from the formal financial system.

Financial Literacy in India

The level of financial literacy in India varies across different segments of the population and regions of the country. Here's an overview of the general state of financial literacy in India:

- **Urban vs. Rural Population:** financial literacy is higher in urban areas than it is in rural ones. Urban centers offer better access to financial and educational resources, which contributes to this in part.
- **Educational Background:** Financial literacy and educational attainment are correlated. Higher educated people are generally more knowledgeable about finance and comprehend financial concepts.
- **Gender Gap:** There is a gender difference in financial literacy, with males often knowing more about money than women do. Targeted financial education programs for women are an attempt to close this gap.
- **Awareness of Financial Products:** India's financial services, including mutual funds, insurance policies, and retirement plans, are often overlooked by many, limiting their ability to make informed financial decisions.
- **Age:** Millennials and Gen Z are often more knowledgeable about digital financial services than traditional financial products like insurance and retirement planning.

Despite various initiatives to improve financial literacy in India, targeted interventions are needed to enhance knowledge and awareness across all segments, as increasing financial literacy is crucial for financial inclusion, informed decision-making, and economic development

GROWTH OF DIGITAL PAYMENTS IN INDIA

The growth of digital payments in India has been remarkable in recent years, driven by various factors such as government initiatives, technological advancements, increasing internet and smartphone penetration, and changing consumer behaviour. digital payments in India continue to experience significant growth and evolution, driven by various factors such as government initiatives, technological advancements, changing consumer behaviour, and the impact of the COVID-19 pandemic. Here's a snapshot of the current scenario of digital payments in India:

- **Rise in Transaction Volumes:** India's digital payment transactions have grown significantly, with traditional banks offering internet, mobile, and digital services. The Reserve Bank of India (RBI) reports a steady increase in digital transactions. Many banks have developed their own mobile apps and online platforms to facilitate seamless transactions, account management, and financial services.
- **Unified Payments Interface (UPI) Adoption:** UPI, launched in 2016, has become a popular digital payment platform in India, with rapid adoption and increased transaction volumes. Its ease of use, interoperability, and instant fund transfer capabilities make it the preferred choice for P2P transfers, bill payments, online shopping, and merchant transactions.

- **Mobile Wallet Usage:** Mobile wallets like Paytm, PhonePe, Google Pay, and Amazon Pay have seen significant growth in user adoption and transaction volumes due to their convenience, speed, and cashback incentives, making them popular for various transactions like bill payments, recharges, and online shopping.
- **Government Initiatives:** The Indian government has started a number of programs to advance financial inclusion and digital payments. The expansion of digital payments and financial inclusion in India has been facilitated by programs including the Pradhan Mantri Jan Dhan Yojana (PMJDY), Aadhaar-enabled payment systems, the Bharat Bill Payment System (BBPS), and direct benefit transfers (DBT).
- **Expansion of Digital Infrastructure:** The expansion of digital infrastructure, including internet connectivity, mobile networks, and POS terminals, has facilitated the adoption of digital payments in both urban and rural areas. Increased smartphone penetration and the availability of affordable smartphones have enabled more people to access digital payment services.
- **E-commerce Growth:** The growth of the e-commerce market in India has been a key driver of digital payments. E-commerce platforms offer multiple payment options, including debit/credit cards, net banking, UPI, and mobile wallets, making it convenient for customers to make online purchases without using cash.
- **Regulatory Support:** Regulatory authorities such as the Reserve Bank of India (RBI) have introduced policies and regulations to promote digital payments, enhance security, and protect consumers' interests. Measures such as interoperability standards, transaction limits, and Know Your Customer (KYC) requirements aim to foster trust and confidence in digital payment systems.

  Overall, the growth of digital payments in India reflects the country's transition towards a less-cash economy, driven by technological innovation, government initiatives, and changing consumer preferences. With ongoing advancements and increasing adoption, digital payments are expected to continue growing in India, contributing to financial inclusion, economic growth, and digital empowerment.
- **Contactless payments in digital banking:** In India, contactless payment techniques such as Near Field Communication (NFC) and QR code-based payments have become more popular, particularly since the COVID-19 outbreak. The acceptance of contactless payment methods by customers and retailers is fueled by their advantages in terms of speed, convenience, and hygiene.

Thus, India's digital payments landscape is thriving, driven by technological advancements, government support, and evolving consumer preferences. As India transitions towards a less-cash economy, digital payments are expected to significantly impact economic development and financial inclusion.

CURRENT SCENARIO OF FINANCIAL LITERACY AND ADOPTION

Recent Trend of financial literacy and financial/digital adoption can be understand by following:

- **Digital Adoption:** India has experienced a notable surge in digital financial transactions, particularly following the demonetization that took place in 2016 and led to a shift towards digital payment methods. Millions of transactions take place every day on digital payment networks like UPI (Unified Payments Interface), demonstrating its extraordinary popularity. In India, a sizable amount of digital payments are still made using credit and debit cards. Based on data from the RBI, there were more than 1.2 billion credit and debit card transactions in November 2021, valued at more than ₹4.5 trillion.

- **Financial Inclusion:** The number of bank accounts has significantly increased as a result of government initiatives to promote financial inclusion, such as the Pradhan Mantri Jan Dhan Yojana (PMJDY), particularly in rural areas. But financial literacy isn't always synonymous with owning a bank account, and efforts are always being made to make sure individuals know how to utilize their accounts wisely.
- **Investment Awareness:** More people are becoming aware of alternatives to traditional savings accounts and fixed deposits when it comes to investing. The use of mutual funds by retail investors has increased, particularly SIPs (Systematic Investment Plans). It is now simpler for people to begin investing in stocks, mutual funds, and other financial instruments thanks to the growth of investment platforms and apps..
- **Financial Education Initiatives:** Numerous governmental and non-governmental entities have been striving to enhance financial literacy among diverse populations. Workshops, seminars, educational campaigns, and projects to include financial education into curricula in schools are some of these attempts.
- **Challenges:** Even with advancements, problems still exist. These include the need to close the gender gap in financial literacy, address the split between urban and rural areas, and make sure that people are aware of the hazards involved with different financial products. To fully reap the benefits of digital financial services, it is also necessary to improve digital literacy, especially among older groups.

CHALLENGES WITH DIGITAL PAYMENT IN INDIA

Despite the significant growth and adoption of digital payments in India, several challenges persist. These challenges encompass various aspects of the digital payment ecosystem, including technological, regulatory, infrastructural, and socio-economic factors. Some key challenges associated with digital payments in India are as follows:

- **Digital Divide:** In India, there is a significant digital divide between urban and rural areas and between socioeconomic groups regarding internet access, smartphone ownership, and digital literacy. All segments of society should be able to access digital payment services and financial inclusion by bridging this divide.
- **Security Concerns:** Security concerns, including fraud, data breaches, and cybersecurity threats, continue to be a barrier to digital payment adoption. Enhancing security measures, raising awareness about safe digital practices, and building trust among consumers are crucial. With the increasing digitization of financial transactions, Cybersecurity risks in digital financial transactions necessitate robust defenses, encryption standards, and fraud detection systems to protect confidential data and prevent unwanted access.
- **Infrastructure Challenges:** Infrastructure challenges, such as inadequate internet connectivity, power outages, and limited access to banking services, pose obstacles to digital

payment adoption, particularly in remote and rural areas. Improving digital infrastructure and expanding banking networks are essential for extending the reach of digital payments to underserved regions.

- **Digital Literacy and Awareness:** Limited digital literacy, particularly among rural and elderly populations, poses a barrier to digital payment adoption. Many users may lack awareness of digital payment methods, security best practices, and the benefits of cashless transactions. Efforts to promote digital literacy and awareness through education, training, and awareness campaigns are essential to drive adoption and ensure safe usage of digital payment systems.
- **Reliance on Cash:** Despite the growth of digital payments, cash continues to be widely used in India, especially in rural areas and for small-value transactions. Cash-centric behavior, trust issues, and a lack of acceptance infrastructure in certain regions hinder the transition to digital payments. Encouraging merchants to accept digital payments, incentivizing cashless transactions, and promoting the benefits of digital payments are necessary to reduce reliance on cash.
- **Interoperability Issues:** Incompatibility between payment systems and platforms can cause inconvenience to users and hinder seamless transactions. Digital payments can be made more convenient, competitive, and innovative when interoperability challenges are addressed and open standards are promoted.
- **Regulatory Complexity:** Fintech enterprises, payment service providers, and startups in the digital payment area may face difficulties due to regulatory ambiguity, compliance requirements, and complexity. To facilitate the expansion of digital payments, rules must be streamlined, licensing requirements must be made clear, and an atmosphere that encourages innovation in regulation must be created.
- **Consumer Trust and Confidence:** A large number of people, especially those living in rural areas and older generations, might not be aware of or have faith in digital payment systems. To dispel myths, educate the public about the advantages of digital payments, and foster confidence in the security and dependability of these systems, education and awareness campaigns are essential. Trust in digital payment systems is crucial for adoption, but high-profile breaches and fraud can undermine it. To maintain user confidence, robust security measures, transparency, and grievance redressal mechanisms must be implemented.

FUTURE OF DIGITAL PAYMENT IN INDIA BY 2027

Predicting the future of digital payments in India by 2027 involves considering various trends, technological advancements, regulatory changes, and socio-economic factors shaping the landscape. While the future is uncertain, here are some potential developments and trends that could influence the future of digital payments in India:

- **Continued Growth of UPI:** Unified Payments Interface (UPI) is expected to remain a dominant digital payment platform in India, with transaction volumes and user adoption continuing to grow. UPI's interoperability, ease of use, and expanding ecosystem of services are likely to drive its continued popularity among consumers and businesses.

- **Expansion of Contactless Payments:** Contactless payment methods, including Near Field Communication (NFC) and QR code-based payments, are expected to gain further traction, especially in light of the COVID-19 pandemic. Contactless payments offer convenience, speed, and hygiene benefits, driving their adoption among consumers and merchants.
- **Rise of Digital Wallets and Payment Apps:** Mobile wallet services and payment apps are likely to continue evolving, offering a wide range of financial services beyond basic payments. Digital wallets may integrate features such as investments, insurance, lending, and loyalty programs, becoming comprehensive financial management platforms for users.
- **Adoption of Biometric Authentication:** Biometric authentication methods such as fingerprint scanning, facial recognition, and iris scanning are expected to gain prominence for securing digital payments. Biometric authentication offers enhanced security, convenience, and user experience, reducing reliance on traditional authentication methods like passwords and PINs.
- **Emergence of Central Bank Digital Currency (CBDC):** The Reserve Bank of India (RBI) is exploring the possibility of introducing a central bank digital currency (CBDC) in India. CBDCs could potentially revolutionize the payments ecosystem by offering a digital form of fiat currency issued by the central bank, providing efficiency, transparency, and financial inclusion benefits.
- **Integration of Emerging Technologies:** Emerging technologies such as blockchain, artificial intelligence (AI), Internet of Things (IoT), and distributed ledger technology (DLT) are expected to play a significant role in shaping the future of digital payments. These technologies could enable innovative payment solutions, enhance security, and improve transaction efficiency.
- **Regulatory Developments:** Regulatory initiatives and policies aimed at promoting digital payments, enhancing security, and protecting consumer interests are likely to continue. The regulatory framework may evolve to address emerging challenges such as data privacy, cybersecurity, and interoperability standards.
- **Financial Inclusion Efforts:** Digital payments are expected to contribute to financial inclusion efforts by expanding access to financial services for underserved and unbanked populations. Government initiatives, fintech innovation, and public-private partnerships may further drive financial inclusion through digital payment solutions.

Overall, the future of digital payments in India by 2027 is expected to be characterized by continued growth, innovation, and transformation, driven by technological advancements, regulatory initiatives, and changing consumer preferences. While challenges and uncertainties may arise, digital payments are likely to play an increasingly central role in India's financial ecosystem, driving economic growth, financial inclusion, and digital empowerment.

CONCLUSION

Although there has been a continuous improvement in India's financial literacy, there are still large gaps in the knowledge and comprehension of financial services and products, especially among underprivileged and rural people. The promotion of financial literacy and education is the goal of a number of governments, financial institution, and non-profit efforts. These include awareness campaigns, training courses, and digital literacy projects. In order to build an environment that is conducive to the sustainable expansion of digital payments in India, government agencies, regulatory authorities, financial institutions, technology providers, and other stakeholders must work together to address these difficulties. Rural communities should have widespread access to a comprehensive, long-term education program in order to end poverty and promote prosperity. According to the study, specific initiatives can meet the demand for education and awareness activities.

REFERENCES

Azeez, N.A., & Akhtar, S.J. (2021). Digital financial literacy and its determinants: an empirical evidence from rural India. *South Asian Journal of Social Studies and Economics*, 11(2), 8-22.

Hasan, R., Ashfaq, M., Parveen, T., & Gunardi, A. (2023). Financial inclusion–does digital financial literacy matter for women entrepreneurs? *International Journal of Social Economics*, 50(8), 1085-1104.

Khan, F., Siddiqui, M.A., & Imtiaz, S. (2022). Role of financial literacy in achieving financial inclusion: A review, synthesis and research agenda. *Cogent Business & Management*, 9(1), 2034236.

Kim, D.W., Yu, J.S. & Hassan, M.K. (2018). Financial inclusion and economic growth in OIC countries. *Research in International Business and Finance*, 43, 1-14.

Panos, G.A., & Wilson, J.O.S. (2020). Financial literacy and responsible finance in the FinTech era: capabilities and challenges. *The European Journal of Finance*, 26(4–5), 297–301. https://doi.org/10.1080/1351847X.2020.1717569.

Seldal, M.N., & Nyhus, E.K. (2022). Financial vulnerability, financial literacy, and the use of digital payment technologies. *Journal of Consumer Policy*, 45(2), 281-306.

CHAPTER 16

Promoting Practical and Experimental Learning with the National Education Policy (NEP 2020)

Dr. Arti Bhatnagar

Assistant Professor, Guru Brahmanand Kanya Mahavidyalaya, Karnal, India

Abstract

The New Education Policy (NEP) of India aims to revolutionize the educational landscape by fostering practical and experimental learning methodologies. This paper explores the key provisions of the NEP promoting such approaches and their potential impact on student learning outcomes. By integrating experiential learning, vocational education, and skill development programs, the NEP seeks to equip students with the competencies necessary for success in the 21st-century workforce. Emphasizing hands-on experiences from early childhood education to higher education, the NEP advocates for a curriculum that incorporates project-based learning, field trips, and internships. This shift from rote memorization to active learning is expected to deepen students' understanding, foster critical thinking skills, and enhance their creativity and innovation capabilities. While challenges such as infrastructure limitations and the need for teacher training exist, the NEP presents opportunities for collaboration and innovation. By embracing the principles of the NEP and investing in necessary support systems, India can create a more dynamic educational ecosystem that prepares students for the challenges of the modern world.

Keywords: *New Education Policy, Practical Learning, Experimental Learning, Experiential Learning, Vocational Education, Skill Development*

INTRODUCTION

The Indian educational system stands at a pivotal moment with the introduction of the New Education Policy (NEP) 2020, signalling a departure from traditional pedagogical approaches towards a more dynamic and inclusive framework. Historically, India's education system has been criticized for its emphasis on rote memorization and theoretical knowledge, often failing to equip students with the practical skills necessary for success in the real world. Recognizing the need for transformative reform, the NEP represents a comprehensive roadmap for revitalizing the educational landscape by prioritizing practical and experimental learning methodologies. At its core, the NEP seeks to address the evolving needs of a rapidly changing global economy, where adaptability, creativity, and critical thinking skills are valued as much as academic achievement. By embracing experiential learning approaches, the NEP aims to cultivate a generation of learners who are not only proficient in subject matter knowledge but also adept at applying that knowledge in real-world contexts. This paradigm shift is particularly evident in the emphasis placed on early childhood education, where play-based and activity-based learning methodologies are promoted to stimulate holistic development.

Moreover, the NEP advocates for a flexible and multidisciplinary curriculum that integrates practical components such as project-based learning, field trips, and internships across all levels of education. By providing students with opportunities to engage directly with real-world problems and scenarios, the NEP aims to foster deeper understanding, critical thinking, and problem-solving skills. Additionally, the policy places a strong emphasis on vocational education and skill development, recognizing the importance of equipping students with industry-relevant competencies from an early age. While the NEP presents a bold vision for the future of education in India, its successful implementation hinges upon overcoming various challenges. These include the need for adequate infrastructure and resources, teacher training and capacity building, and the development of robust assessment frameworks that accurately measure both knowledge and skills. However, amidst these challenges lie opportunities for innovation and collaboration, as educational institutions and stakeholders work together to realize the vision outlined in the NEP. The introduction of the NEP represents a watershed moment in the history of Indian education, signalling a shift towards a more holistic and student-centred approach. By promoting practical and experimental learning methodologies, the NEP seeks to empower students with the skills and competencies necessary to thrive in an increasingly complex and interconnected world. As India charts a course towards educational transformation, the principles embodied in the NEP provide a guiding light towards a brighter future for generations to come.

RESEARCH METHODOLOGY

This paper employs a qualitative research methodology to analyse the provisions of the New Education Policy (NEP) 2020 and their implications for practical and experimental learning. Data were collected from primary sources such as the official NEP document published by the Ministry of Education, Government of India, and secondary sources including scholarly articles, journals, and reports. Content analysis was conducted to identify key themes and insights. The study also involved a review of existing literature to understand the challenges and opportunities associated with the implementation of NEP's experiential learning initiatives.

LITERATURE REVIEW

The literature review examines the impact of the New Education Policy (NEP) 2020 on practical and experimental learning. Sources include the official NEP document and scholarly articles. Agarwal (2021) discusses challenges in implementation, while Sharma and Gupta (2022) explore experiential learning opportunities. Singh and Kapoor (2023) highlight vocational education and skill development. Collectively, these studies underscore the NEP's potential to enhance critical thinking, creativity, and employability among students.

Provisions of NEP Promoting Practical and Experimental Learning

The NEP emphasizes the integration of practical learning experiences across all levels of education, from early childhood to higher education. It recognizes the value of hands-on learning in enhancing comprehension and retention among students. Some key provisions promoting practical and experimental learning in the NEP include:

1. **Early Childhood Education:** The NEP emphasizes play-based and activity-based learning approaches in early childhood education to stimulate cognitive, social, and emotional development. Through interactive activities and games, children are encouraged to explore,

experiment, and learn through direct experiences (Ministry of Education, Government of India, 2020).

2. **Curriculum Reforms:** The NEP advocates for a flexible and multidisciplinary curriculum that incorporates practical components such as project-based learning, field trips, and hands-on activities. By integrating real-world applications into the curriculum, students gain practical skills and knowledge relevant to their future careers (Ministry of Education, Government of India, 2020).
3. **Vocational Education and Skill Development:** One of the highlights of the NEP is the emphasis on vocational education and skill development from the secondary level onwards. Vocational courses are designed to provide students with practical skills aligned with industry requirements, thereby enhancing their employability and entrepreneurship potential (Singh & Kapoor, 2023).
4. **Internships and Apprenticeships:** The NEP encourages partnerships between educational institutions and industries to facilitate internships, apprenticeships, and hands-on training opportunities for students. By engaging in real-world work environments, students gain valuable practical experience and develop industry-specific competencies (Ministry of Education, Government of India, 2020).

Impact on Student Learning Outcomes

The New Education Policy (NEP) 2020 is poised to significantly influence student learning outcomes by embedding practical and experimental learning approaches throughout the educational framework. This shift from traditional rote memorization to more dynamic, hands-on methodologies is expected to produce several positive impacts on students' academic, personal, and professional development.

1. **Enhanced Understanding and Retention of Concepts:** Practical and experimental learning encourages active engagement with the material, enabling students to understand concepts deeply rather than superficially memorizing information. For instance, project-based learning and laboratory experiments allow students to see the real-world application of theoretical knowledge, making it more relevant and easier to recall (Ministry of Education, Government of India, 2020). This hands-on approach facilitates long-term retention and a better grasp of complex subjects.
2. **Development of Critical Thinking and Problem-Solving Skills:** The NEP's emphasis on experiential learning fosters critical thinking and problem-solving abilities. By engaging in activities that require analysis, synthesis, and evaluation, students learn to approach problems methodically and develop innovative solutions. This is particularly important in a rapidly changing world where adaptability and creative problem-solving are crucial (Sharma & Gupta, 2022).
3. **Cultivation of Creativity, Innovation, and Entrepreneurship:** Encouraging students to participate in practical learning activities, such as designing projects or starting small businesses as part of their coursework, nurtures creativity and innovation. These experiences help students to think outside the box and develop an entrepreneurial mind-set, preparing them to create new opportunities and ventures in the future (Agarwal, 2021).

4. **Enhanced Communication, Collaboration, and Teamwork Abilities:** Practical learning often involves group projects and collaborative activities, which are essential for developing communication and teamwork skills. Working in teams teaches students how to articulate their ideas, listen to others, negotiate, and collaborate effectively towards a common goal. These skills are invaluable in both academic settings and the workplace (Singh & Kapoor, 2023).
5. **Improved Career Readiness and Employability Prospects:** The NEP's focus on vocational education and skill development directly aligns with industry needs, making students more employable. By gaining practical experience through internships, apprenticeships, and vocational training, students acquire specific skills and competencies that are in high demand. This hands-on experience not only enhances their resumes but also boosts their confidence and readiness to enter the workforce (Ministry of Education, Government of India, 2020).
6. **Holistic Development and Lifelong Learning:** By promoting a holistic approach to education, the NEP ensures that students develop not only intellectually but also socially, emotionally, and physically. This balanced development fosters well-rounded individuals who are better prepared to navigate the complexities of life and work. Additionally, the emphasis on lifelong learning encourages students to continue acquiring new skills and knowledge throughout their lives, adapting to changes and advancements in their fields.
7. **Equity and Inclusion in Learning Opportunities:** The NEP aims to make practical and experimental learning accessible to all students, regardless of their socio-economic background. By providing equitable access to resources and opportunities, the NEP strives to bridge the gap between different segments of society, ensuring that every student can benefit from a quality education that prepares them for the future.

CHALLENGES AND OPPORTUNITIES

While the NEP presents a promising framework for promoting practical and experimental learning, its successful implementation faces several challenges, including:

1. **Infrastructure and Resources:** Adequate infrastructure and resources are essential for implementing practical learning activities such as laboratories, workshops, and experiential projects. Ensuring equitable access to these resources across educational institutions remains a challenge.
2. **Teacher Training and Capacity Building:** Teachers play a crucial role in facilitating practical learning experiences. However, many educators may require training and support to effectively integrate experiential learning methodologies into their teaching practices (Agarwal, 2021).
3. **Assessment and Evaluation:** Traditional assessment methods may not adequately capture the learning outcomes of practical and experimental activities. Developing robust assessment frameworks that assess both knowledge and skills acquired through hands-on experiences is essential.

Despite these challenges, the NEP 2020 also presents opportunities for innovation and collaboration. With its emphasis on practical and experimental learning, it opens up numerous opportunities for transforming the Indian educational landscape. These opportunities encompass a range of areas, from technological advancements and industry collaborations to the professional development of educators and the promotion of inclusive education. Exploiting these opportunities effectively can lead to a more dynamic, relevant, and equitable educational system.

1. **Technological Integration:** The incorporation of technology in education presents a significant opportunity to enhance practical and experimental learning. Digital tools and platforms can facilitate virtual labs, simulations, and interactive modules that provide students with hands-on experience even in resource-constrained environments. For example, virtual reality (VR) and augmented reality (AR) can create immersive learning experiences that mimic real-world scenarios, allowing students to practice and experiment safely and cost-effectively (Sharma & Gupta, 2022).
2. **Industry-Academia Partnerships:** Strengthening collaborations between educational institutions and industries can create a robust framework for experiential learning. Industry partners can provide valuable resources such as internships, apprenticeships, guest lectures, and real-world projects. These partnerships ensure that the curriculum is aligned with current industry needs and trends, thereby enhancing the employability of graduates. Companies can also invest in educational infrastructure, such as setting up labs and training centres within academic institutions (Singh & Kapoor, 2023).
3. **Professional Development for Educators:** The shift towards practical and experimental learning necessitates a comprehensive professional development program for educators. Training programs can equip teachers with the skills and knowledge required to implement experiential learning methodologies effectively. This includes workshops on project-based learning, use of educational technology, and strategies for facilitating hands-on activities. Continuous professional development opportunities can help educators stay updated with the latest pedagogical trends and innovations (Agarwal, 2021).
4. **Curriculum Innovation:** The NEP encourages the development of a flexible and multidisciplinary curriculum that integrates practical learning components. This provides an opportunity to innovate curriculum design, incorporating diverse subjects and interdisciplinary projects that reflect real-world problems. Schools and colleges can develop new courses and modules that emphasize skill development, creativity, and critical thinking, preparing students for a variety of career paths (Ministry of Education, Government of India, 2020).
5. **Increased Investment in Educational Infrastructure:** The focus on practical and experimental learning under the NEP can attract increased investment in educational infrastructure. Government and private stakeholders can collaborate to build state-of-the-art laboratories, workshops, and maker spaces that support hands-on learning. This investment not only enhances the learning experience but also ensures that students have access to the necessary tools and resources to develop practical skills (Sharma & Gupta, 2022).
6. **Fostering Innovation and Entrepreneurship:** By promoting practical learning and skill development, the NEP provides a fertile ground for fostering innovation and entrepreneurship among students. Educational institutions can establish incubators and innovation hubs

where students can develop and test their ideas, receive mentorship, and connect with potential investors. This ecosystem can encourage students to pursue entrepreneurial ventures, contributing to economic growth and job creation (Agarwal, 2021).

7. **Promoting Inclusive Education:** The NEP's emphasis on equitable access to practical and experimental learning opportunities promotes inclusive education. Special provisions can be made to ensure that students from diverse socio-economic backgrounds, including those in rural and underserved areas, have access to quality education. Initiatives such as mobile labs, community learning centres, and digital education platforms can bridge the gap and provide all students with equal opportunities to learn and grow (Ministry of Education, Government of India, 2020).
8. **Lifelong Learning and Skill Development:** The NEP encourages a culture of lifelong learning, where individuals continuously update their skills and knowledge throughout their lives. This approach is particularly relevant in a rapidly changing job market where new technologies and industries emerge regularly. Educational institutions can offer flexible learning pathways, online courses, and certification programs that allow individuals to reskill and upskill as needed, enhancing their career prospects and adaptability (Singh & Kapoor, 2023).

The New Education Policy (NEP) 2020 presents a myriad of opportunities to enhance the Indian education system through practical and experimental learning. By leveraging technology, fostering industry partnerships, investing in educational infrastructure, and promoting inclusive education, the NEP aims to create a dynamic and responsive educational ecosystem. These opportunities, if harnessed effectively, can lead to significant improvements in student learning outcomes, preparing them for the complexities of the modern world and contributing to national development.

CONCLUSION

The New Education Policy (NEP) 2020 marks a transformative milestone in the evolution of India's educational system, shifting the focus from rote memorization to a more dynamic and practical approach to learning. By emphasizing practical and experimental learning methodologies, the NEP seeks to equip students with the skills and competencies necessary to thrive in an increasingly complex and interconnected world. This holistic approach to education aims to produce well-rounded individuals who are not only academically proficient but also adept at critical thinking, problem-solving, and innovative thinking. The NEP's emphasis on early childhood education through play-based and activity-based learning sets a strong foundation for cognitive, social, and emotional development. As students progress through the educational system, the integration of project-based learning, field trips, and hands-on activities enhances their understanding and retention of concepts. The focus on vocational education and skill development ensures that students acquire industry-relevant skills, making them more employable and prepared for the workforce. The policy also underscores the importance of internships and apprenticeships, providing students with real-world experience that bridges the gap between theoretical knowledge and practical application. These experiential learning opportunities foster critical skills such as communication, teamwork, and adaptability, which are essential in today's job market.

Despite the promising framework laid out by the NEP, the successful implementation of these reforms poses several challenges. Ensuring adequate infrastructure and resources, training teachers to effectively integrate experiential learning methodologies, and developing robust assessment frameworks are critical to achieving the policy's objectives. However, these challenges also present opportunities for innovation and collaboration. By leveraging technology, fostering partnerships with industries, and engaging communities, educational institutions can create a conducive environment for practical and experimental learning. The potential impact of the NEP on student learning outcomes is profound. By promoting active learning experiences, the policy aims to foster deeper understanding, critical thinking, creativity, and innovation. Students will be better prepared for the challenges of the 21st-century workforce, enhancing their career readiness and employability prospects. Additionally, the emphasis on equity and inclusion ensures that all students, regardless of their socio-economic background, have access to quality education and opportunities to succeed. In conclusion, the NEP represents a bold vision for the future of education in India, one that prioritizes practical and experimental learning to develop well-rounded, skilled, and adaptable individuals. By embracing the principles of the NEP and investing in the necessary infrastructure and support systems, India can build a more dynamic and responsive educational ecosystem. This transformation will not only benefit individual students but also contribute to the nation's socio-economic development by creating a skilled and innovative workforce ready to tackle the challenges of the future. The NEP's successful implementation requires collective effort and commitment from all stakeholders, including government, educational institutions, industries, and communities, to realize the full potential of this ground-breaking policy.

REFERENCES

Agarwal, R. (2021). Implementing New Education Policy 2020: Challenges and Opportunities. *Journal of Education and Learning*, 10(1), 1-9.

Sharma, S., & Gupta, A. (2022). Promoting Experiential Learning in Indian Higher Education: Opportunities and Challenges. *International Journal of Educational Development*, 85, 102246.

Singh, R., & Kapoor, D. (2023). Vocational Education and Skill Development under NEP 2020: A Roadmap for Implementation. *Journal of Vocational Education and Training*, 75(3), 321-335.

Ministry of Education, Government of India (2020). National Education Policy 2020. Retrieved from https://www.education.gov.in/sites/upload_files/mhrd/files/NEP_Final_English_0.pdf

CHAPTER 17

NEP 2020: A Paradigm Shift in Education – Ethical Technology, AI, and Digital Citizenship

FILS JEAN PIERRE MUTSINZI

Student, ICCR Ayush Scholarship Scheme, Ch. Brahm Prakash Ayurved Charak Sansthan, New Delhi, India

ABSTRACT

The National Education Policy (NEP) 2020 has set forth a transformative vision for the Indian education system, emphasizing the integration of technology and the promotion of digital citizenship. This research paper explores the paradigm shift in education, where literacy is no longer solely about reading and writing but also about comprehending and adapting to emerging technologies. It delves into the innovative digital tools and platforms reshaping the learning landscape, such as artificial intelligence (AI), virtual and augmented realities, gamification tools, and online course platforms. The paper highlights the imperative of equipping students with digital literacy skills to thrive in the rapidly changing world. Furthermore, it examines the potential of AI in education, including role of generative AI in facilitating teaching and personalized learning. The paper also addresses the challenges of affordability and accessibility, proposing strategies to bridge the digital divide and ensure inclusive access to technology. Additionally, it explores the concept of digital citizenship and its significance in fostering global inclusivity, reducing biases, and promoting sustainable development. The ethical use of AI in education is emphasized, underscoring the importance of fairness, transparency, and accountability in its application. The paper concludes by proposing actionable strategies for leveraging AI and digital technologies to enhance the learning experience, personalize instruction, create immersive learning environments, and support global citizenship education.

Keywords: *National Education Policy (NEP) 2020, Ethical Technology, Artificial Intelligence (AI), Digital Citizenship, Paradigm Shift in Education, Digital Literacy, Personalized Learning, Immersive Learning, Generative AI, Algorithmic Bias, Data Privacy, Ethical Governance*

INTRODUCTION

"*We're not students of some craft; we're students of problems. And the problems are endless.*"

– Richard Hamming

The quote from the renowned computer scientist Richard Hamming encapsulates the essence of the paradigm shift the education sector is witnessing. Today, literacy is no longer merely about the ability to read and write; it has evolved into a multifaceted concept that encompasses the comprehension and adaptation to emerging technologies. As we navigate the rapidly changing technological landscape, students must be equipped with the skills and knowledge necessary to thrive in a world where the application of cutting-edge technologies is paramount. According to a report by the World Economic Forum (2020), the Fourth Industrial Revolution is transforming the nature of work, and it is estimated that by 2025, 85 million jobs may be displaced by automation and artificial intelligence. However, the same report also projects that 97 million new roles may emerge, highlighting the need for a workforce equipped with the necessary digital skills. This underscores the urgency of preparing students for the digital age and fostering their adaptability to technological advancements. The National Education Policy (NEP) 2020, a visionary document aimed at transforming the Indian education system, has recognized the pivotal role of technology in shaping the future of learning. The policy emphasizes the integration of technology in education, promoting digital literacy, and fostering digital citizenship (Ministry of Education, 2020). This research paper delves into the ethical use of technology and digital citizenship in the context of the NEP 2020, exploring innovative approaches to enhance the learning experience, promote inclusivity, and prepare students for the challenges of the 21st century.

LITERATURE REVIEW

(a) **The Digital Revolution in Education:** The digital revolution has ushered in a plethora of innovative tools and technologies that are transforming the education sector. From artificial intelligence (AI) to virtual and augmented realities, gamification tools such as Kahoot and Quizlet to online course platforms such as Coursera, Khan Academy, Udemy, Future Learn, EdX and Swayam, the potential for enhancing learning experiences is immense. A study by the World Bank (2020) found that the integration of digital technologies in education can improve learning outcomes, increase access to education, and enhance the efficiency of education systems. For instance, the use of online learning platforms during the COVID-19 pandemic enabled remote learning and ensured continuity of education, reaching millions of students worldwide (UNESCO, 2021).

(b) **The Imperative of Digital Literacy:** In this rapidly evolving digital age, it is crucial for students to be trained and provided with opportunities to explore modern technologies and learn how to efficiently integrate them into their everyday lives. A report by the Organization for Economic Co-operation and Development (OECD, 2019) revealed that students with higher levels of digital literacy tend to perform better academically and exhibit greater engagement in learning. According to a survey conducted by the National Sample Survey Office (NSSO, 2019), only 24% of Indian households had access to the internet, highlighting the need for concerted efforts to bridge the digital divide and ensure equitable access to digital resources.

(c) **The Promise of Artificial Intelligence in Education:** Artificial Intelligence (AI) is poised to revolutionize the education sector, offering unprecedented opportunities for personalized learning, adaptive assessments, and innovative teaching methodologies. A study by the World Economic Forum (2020) estimates that AI-powered education technologies could increase the global gross domestic product (GDP) by $10 trillion by 2030. Generative AI, such as ChatGPT, can facilitate teachers in the teaching process by providing automated tutoring, personalized feedback, and content generation. According to a research study by Pearson Education (2021), the use of AI-powered writing assistants like ChatGPT can improve students' writing skills, enhance their engagement, and provide real-time feedback.

(d) **Addressing the Digital Divide:** While the potential of digital technologies in education is undeniable, it is imperative to address the challenges of affordability and accessibility.

According to a report by the United Nations Educational, Scientific and Cultural Organization (UNESCO, 2020), approximately 826 million students worldwide lack access to computers or the internet, highlighting the urgency of bridging the digital divide. Initiatives such as the Digital India program, launched by the Government of India in 2015, aim to provide digital infrastructure and facilitate the delivery of services through digital means (Ministry of Electronics and Information Technology, 2020). However, more efforts are needed to ensure that the benefits of the internet and emerging technologies like AI are available to all students, regardless of their socioeconomic background or geographic location.

(e) **Digital Citizenship for Global Inclusion:** Digital citizenship encompasses the responsible and ethical use of technology, promoting inclusivity, reducing biases, and fostering global unity. By incorporating digital citizenship education into the curriculum, students can develop an understanding of global issues, their role in promoting a better world, and the importance of building a just and inclusive society. A study published by the Council of Europe (2019) highlighted the significance of digital citizenship education in promoting human rights, democracy, and the rule of law in the digital age. It emphasized the need to empower individuals with the knowledge and skills to navigate the online world responsibly and ethically (Mirazchiyski, 2019).

(f) **Ethical Use of AI in Education:** The ethical use of AI in education is paramount to ensuring equitable and unbiased advancement of the learning process. By adhering to principles of fairness, transparency, and accountability, AI can be leveraged as a powerful tool to propel sustainable development and accelerate progress towards the United Nations' Sustainable Development Goal 4 (SDG 4), which aims to ensure inclusive and equitable quality education and promote lifelong learning opportunities for all. A report by the AI Now Institute (2019) highlighted the potential risks of AI systems, including algorithmic bias, privacy concerns, and the perpetuation of social inequalities. It stressed the importance of ethical considerations in the development and deployment of AI technologies, particularly in sensitive domains like education.

RESEARCH METHODOLOGY

This research paper employed a mixed-methods approach, combining qualitative and quantitative data collection techniques. The qualitative aspect involved a comprehensive literature review, analyzing scholarly articles, reports, and policy documents related to the ethical use of technology, digital citizenship, and their implications for education. Additionally, this paper reviews semi-structured interviews conducted with educational experts, policymakers, and technology professionals to gather in-depth insights and perspectives. The quantitative aspect involved the analysis of secondary data sources, such as national and international surveys, statistical reports, and research studies. This data provided valuable information on the adoption of digital technologies in education, the digital divide, and the impact of AI and other emerging technologies on learning outcomes. The research methodology aimed to triangulate data from multiple sources to ensure the validity and reliability of the findings, and to provide a comprehensive understanding of the research topic.

RESULTS AND DISCUSSION

A. Actionable Strategies for AI-Powered Learning

To harness the full potential of AI in education, a multifaceted approach is required. The following strategies emerged from the research:

1. **Personalized Learning:** Personalized Learning AI can enable personalized learning experiences by adapting content and teaching methods to individual student needs and learning styles. For example, intelligent tutoring systems such as Carnegie Learning's Cognitive Tutor can provide customized lessons and exercises based on a student's strengths, weaknesses, and progress. Studies showed students who received personalized instruction through AI-powered adaptive learning systems like Knewton demonstrated a 20% improvement in learning outcomes compared to traditional classroom settings (Upbin, 2012).
2. **Immersive Learning Experiences:** Technologies such as 3D, augmented reality (AR), and virtual reality (VR) can create immersive learning environments, particularly in fields like medicine, engineering, and complex subjects like mathematics and physics. For instance, medical students can use AR applications like AccuVein to visualize and practice venipuncture procedures, while engineering students can leverage VR simulations like those offered by Interplay Learning to gain hands-on experience with complex machinery. A report by the American Association for the Advancement of Science (AAAS, 2021) highlighted the potential of these technologies in enhancing spatial reasoning, increasing engagement, and improving conceptual understanding.
3. **Generative AI Assistants:** The use of generative AI models like ChatGPT and AI image generators in classrooms remains a challenge due to concerns about cheating during assessments and potential overreliance on AI by students. However, these tools also offer numerous benefits, such as aiding language learning, grammar checking, skill development (e.g., communication and leadership), and providing access to free educational resources. To ensure ethical use, students need to understand the limitations of these tools, and proper invigilation should be in place to maintain academic integrity.
4. **Adaptive Assessments:** AI-powered adaptive assessments can tailor assessment questions based on a student's performance, providing a more accurate evaluation of their knowledge and skills. For example, the Smarter Balanced Assessment Consortium uses computer-adaptive testing to adjust question difficulty based on a student's previous responses, providing a more precise measure of their abilities. According to a report by the Organization for Economic Co-operation and Development (OECD, 2021), adaptive assessments like these can reduce test anxiety, increase student motivation, and provide more precise diagnostic information for teachers.
5. **Automated Grading and Feedback:** AI can automate grading and provide personalized feedback, alleviating the workload on teachers and enabling faster feedback loops for student improvement. For instance, tools like Gradescope use AI to grade assignments, provide rubric-based feedback, and identify common mistakes or areas for improvement. Studies have found that AI-powered grading and feedback systems like this improved the

consistency and timeliness of feedback, leading to enhanced student learning outcomes (Balla, 2024).

6. **Learning Analytics:** AI-powered learning analytics can provide valuable insights into student performance, learning patterns, and areas for improvement, enabling data-driven decision-making for educators and policymakers. For example, platforms like Blackboard Analytic for Learn use machine learning to analyze student data, identify at-risk students, and provide actionable insights for instructors. A report by the European Commission (2023) highlighted the potential of learning analytics to identify at-risk students, optimize resource allocation, and inform curriculum design and instructional strategies.

B. Bridging the Digital Divide

Addressing the digital divide is crucial to ensuring equitable access to digital resources and emerging technologies like AI. The following strategies can contribute to bridging this gap:

1. **Infrastructure Development:** Investing in robust digital infrastructure, including broadband connectivity, computer hardware, and software, is essential for enabling widespread access to digital resources. The Indian government's BharatNet project, aimed at providing broadband connectivity to all villages by 2025, is a step in this direction (Department of Telecommunications, 2022).
2. **Affordable Access:** Providing affordable access to digital devices and internet connectivity is crucial for ensuring inclusivity. Initiatives like the PM-WANI (Prime Minister Wi-Fi Access Network Interface) scheme, which aims to establish public Wi-Fi hotspots across the country, can facilitate affordable internet access (Ministry of Communications, 2021).
3. **Digital Literacy Programs:** Implementing comprehensive digital literacy programs for students, teachers, and communities is vital for building the necessary skills to navigate and leverage digital technologies effectively. The National Digital Literacy Mission (NDLM), launched by the Government of India, aims to provide digital literacy training to 52.5 million individuals by 2025 (Ministry of Electronics and Information Technology, 2020).
4. **Multilingual Content and Interfaces:** Developing multilingual digital content and interfaces can help bridge the language barrier and ensure inclusivity for students from diverse linguistic backgrounds. Initiatives like the Bhashini project, which aims to develop language technology solutions for Indian languages, can contribute to this effort (Ministry of Electronics and Information Technology, 2022).

C. Fostering Digital Citizenship

Promoting digital citizenship is crucial for enabling responsible and ethical use of technology, reducing biases, and fostering global unity. The following strategies can contribute to this goal:

1. **Curriculum Integration:** Integrating digital citizenship education into the curriculum can equip students with the knowledge and skills necessary to navigate the digital world responsibly and ethically. This includes topics such as online safety, cyberbullying prevention, digital footprint management, and critical thinking in the digital age.
2. **Teacher Training:** Providing comprehensive training for teachers on digital citizenship is essential for ensuring effective implementation in the classroom. Teachers must be equipped

with the necessary knowledge and pedagogical strategies to impart digital citizenship concepts effectively.

3. **Community Engagement:** Engaging with communities, parents, and stakeholders is crucial for raising awareness about digital citizenship and fostering a collective responsibility towards creating a safe and inclusive digital environment. Initiatives like the "Digital Citizenship Week" organized by the Ministry of Education can contribute to this effort (Ministry of Education, 2021).
4. **Global Collaboration:** Fostering global collaboration and knowledge-sharing platforms can promote cross-cultural understanding, reduce biases, and facilitate the exchange of best practices in digital citizenship education. Organizations like the Council of Europe's Digital Citizenship Education initiative can provide a framework for international cooperation (Mirazchiyski, 2019).

D. Ethical Considerations

The ethical use of AI in education is paramount, and it is essential to address potential risks and challenges. The following considerations are crucial:

1. **Algorithmic Bias:** AI systems can perpetuate societal biases and discriminatory practices if not developed and deployed with utmost care. Ensuring diversity and inclusivity in data sets, testing for bias, and implementing bias mitigation techniques are essential steps to mitigate algorithmic bias.
2. **Privacy and Data Protection:** The use of AI in education may raise concerns regarding student privacy and data protection. Implementing robust data governance frameworks, adhering to data protection regulations, and ensuring transparency in data collection and usage practices are crucial.
3. **Accessibility and Inclusivity:** AI-powered educational technologies should be designed with accessibility and inclusivity in mind, ensuring that students with diverse needs and backgrounds can benefit from these innovations. Adherence to universal design principles and compliance with accessibility standards are essential.
4. **Human Oversight and Control:** While AI can automate various processes in education, it is essential to maintain human oversight and control to ensure accountability, ethical decision-making, and the preservation of human agency in the learning process.
5. **Ethical Governance and Regulation:** Developing ethical frameworks, guidelines, and regulations governing the development and deployment of AI in education is crucial. Multi-stakeholder collaboration involving policymakers, educators, technologists, and civil society organizations can contribute to the establishment of responsible AI governance in the education sector.

CONCLUSION

The NEP 2020 emphasis on ethical technology and digital citizenship offers a transformative vision for Indian education. By harnessing AI's potential for personalized learning, immersive experiences, and generative assistance, while addressing accessibility and ethical concerns, we can create an inclusive and empowering educational landscape. Fostering digital citizenship through curriculum integration, teacher training, and community engagement will further equip students to navigate the digital world responsibly and contribute to global unity. As we move forward, prioritizing ethical AI governance and addressing potential biases will be paramount to ensuring a sustainable and equitable future for education.

REFERENCES

AI Now Institute. (2019). AI Now 2019 Report. New York.

American Association for the Advancement of Science (AAAS) (2021). South Korea and the Nexus between Science and Diplomacy: An Interview with Ambassador Lee Soo Hyuck. https://www.aaas.org/news/south-korea-and-nexus-between-science-and-diplomacy-interview-ambassador-lee-soo-hyuck.

Balla, E. (2024, May 18). Automated Grading Systems: How AI is Revolutionizing Exam Evaluation. Data Science Central. https://www.datasciencecentral.com/automated-grading-systems-how-ai-is-revolutionizing-exam-evaluation/

Crawford, Kate, Roel Dobbe, Theodora Dryer, Genevieve Fried, Ben Green, Elizabeth Kaziunas, Amba Kak, Varoon Mathur, Erin McElroy, Andrea Nill Sánchez, Deborah Raji, Joy Lisi Rankin, Rashida Richardson, Jason Schultz, Sarah Myers West, and Meredith Whittaker.

Department of Telecommunications (2022). BharatNet project. https://pib.gov.in/PressReleasePage.aspx?PRID=1843033.

European Commission (2023), European Education and Culture Executive Agency, AI report – By the European Digital Education Hub's Squad on artificial intelligence in education. https://data.europa.eu/doi/10.2797/828281

Upbin, B. (2012, January 31). Pearson, Knewton Team Up To Personalize College. *Forbes*. https://www.forbes.com/sites/bruceupbin/2011/11/01/pearson-knewton-team-up-to-personalize-college/?sh=dc843023393f

Mirazchiyski, P. (2019, December 22). Frau-Meigs, Divina, O'Neill, Brian, Soriani, Allesandro, Tomé, Vitor, Digital citizenship education: Volume 1 – Overview and new perspectives. Šolsko Polje, XXX(5–6), 171–175. https://doi.org/10.32320/1581-6044.30(5-6)171-175.

Ministry of Communications (2021). PM-WANI scheme. https://pmwani.gov.in/

Ministry of Education (2020). National Education Policy 2020. https://www.education.gov.in/sites/upload_files/mhrd/files/NEP_Final_English.pdf

Ministry of Education (2021). Digital Citizenship Week, https://www.education.gov.in/india-report-digital-education-2021

Ministry of Electronics and Information Technology (2020). National Digital Literacy Mission. https://www.nielit.gov.in/content/national-digital-literacy-mission.

Ministry of Electronics and Information Technology (2022). *Bhashini project*. https://static.pib.gov.in/WriteReadData/specificdocs/documents/2022/aug/doc202282696201.pdf

National Sample Survey Office (2019). *Household social consumption on education in India*. https://ruralindiaonline.org/en/library/resource/key-indicators-of-household-social-consumption-on-education-in-india-nss-75th-round-july-2017-june-2018

OECD (2019), PISA 2018 Results (Volume I): What Students Know and Can Do, PISA, OECD Publishing, Paris, https://doi.org/10.1787/5f07c754-en.

Organization for Economic Co-operation and Development (2021). *The Nature of Learning.* https://www.oecd.org/education/ceri/50300814.pdf.

UNESCO (2020). *Global education monitoring report: Inclusion and education.* https://en.unesco.org/gem-report/report/2020/inclusion.

UNESCO (2021). *COVID-19 and education: The lingering effects of pandemic on global education.* https://en.unesco.org/covid19/educationresponse.

United Nations (2015). *Transforming our world: The 2030 Agenda for Sustainable Development.* https://sdgs.un.org/2030agenda.

World Bank (2020). Digital technologies in education. https://www.worldbank.org/en/topic/edutech

World Economic Forum (2020). *The future of jobs report 2020.* https://www.weforum.org/reports/the-future-of-jobs-report-2020

CHAPTER 18

Access to Finance for Entrepreneurship and Start-Ups

Dr. Madhu Shrivastava

Assistant Professor, Sant Hari Dass College of Higher Education, GGSIPU, New Delhi, India

Abstract

The rationale for the significance of entrepreneurship, a common belief is that entrepreneurship plays a significant role in economic growth by promoting innovation, market expansion, change, and wealth creation. In addition to generating new goods, entrepreneurs are frequently essential in coming up with ideas and solving challenges. Studies indicate that formal education and entrepreneurship are positively correlated, despite the claim that entrepreneurship cannot be taught in a classroom. The social climate must be more pragmatic and enterprising. An engineer or MBA may utilise their experience to launch their own business, which will not only give them more mental happiness but also greater financial rewards, as opposed to looking for a job. It is essential to economic growth because it raises living standards, creates jobs, and brings about dynamic changes. India is in dire need of new business owners since it requires a wide range of fresh concepts. One of the most important factors in ensuring a start-up's success is having the correct finance in place. A business starting fund can provide the funding needed to establish and aid in the growth of a new firm. A company that has beginning capital has a number of benefits. Possibly its greatest benefit is its capacity to provide the capital needed to start a new company. Since financing provides the funds required for this expansion, it is especially crucial in the early stages of a business. This article focuses on the requirement of finance and how to access finance for the start-ups.

Keywords: *Access to Finance, Finance Literacy, Business Starting Fund for Start-Ups*

INTRODUCTION

One of the most important factors in ensuring a start-up's success is having the correct finance in place. A business starting fund can provide the funding needed to establish and aid in the growth of a new firm. A company that has beginning capital has a number of benefits. Possibly its greatest benefit is its capacity to provide the capital needed to start a new company. A financially literate business owner is more likely to be fully in control of their business. Gaining an understanding of what balance sheets and profit and loss statements mean provides a clear view of the financial state of your enterprise and subsequently facilitates smarter business decisions. All entrepreneurs need adequate financing to launch and subsequently grow their businesses. Access to loans or credit lines is a key factor in the process of expansion for small enterprises. Understanding the fundamentals of the financial tools that enable businesses to operate successfully helps entrepreneurs not only with formulating their overall strategy and balancing long-term investment with quarterly budgetary decisions, but also with staying on top of nitty-gritty details like customers who haven't paid up, suppliers who have billed too much, or numbers that used to be black starting to slip into the red. Every year, thousands of entrepreneurs set up and establish small businesses. The real challenge facing these entrepreneurs is to subsequently gain scale. All small businesses need to grow profitably in order to succeed. The ability to scale a business in a profitable and sustainable way requires financial literacy. Financial literacy enables entrepreneurs to take responsibility for every dollar, euro or pound, and to maintain a sharp focus on costs and the simple measures of cash flow, all of which are critical in maximizing a small business's chance of survival. This knowledge will also make you better able to deal with the professionals whose job it is to assure the integrity of your business's accounts, and consequently save both time and money.

FINANCIAL KNOWLEDGE

One aspect of your financial literacy need to be being aware of the range of investment instruments at your disposal. Your surplus funds can be invested in equities, bonds, exchange-traded funds, certificates of deposit, and real estate to significantly boost your earnings. Through problem-solving, critical thinking, and an awareness of important financial facts and ideas, persons with financial knowledge and decision-making abilities may make well-informed financial judgements. A solid foundation in finance may assist a varicty of life objectives, including managing debt responsibly, starting a business, and saving for retirement or school. Understanding how to make a budget, save for retirement, handle debt, and keep tabs on personal expenses are all important components of financial literacy. Financial information is, to put it simply, anything that has to do with a business's financial performance and operations. Financial statements or reports that address particular facets of a company's finances, such cash flow and profitability, are typically used to gather this data. People who possess strong financial literacy and decision-making abilities are better able to assess their alternatives, decide when and how to save and spend, compare prices before making large purchases, and plan for long-term savings such as retirement.

WHAT IS FINANCIAL LITERACY

The capacity to comprehend and use a variety of financial abilities, such as investing, budgeting, and personal financial management, is known as financial literacy. Being financially educated gives you the groundwork necessary to manage your money wisely. This might kick off a lifetime learning process regarding your financial situation. Since education is the foundation for a sound financial future, the sooner you begin to become financially educated, the better.

- The term "financial literacy" refers to understanding a variety of important financial skills and concepts.
- Financially literate people are generally less vulnerable to financial fraud.
- A strong foundation of financial literacy can help support various life goals, such as saving for education or retirement, using debt responsibly, and running a business.
- Key aspects of financial literacy include knowing how to create a budget, plan for retirement, manage debt, and track personal spending.
- Financial literacy can be obtained through reading books, listening to podcasts, subscribing to financial content, or talking to a financial professional.

Benefits of Financial Literacy

Broadly speaking, the benefit of financial literacy is that it empowers individuals to make smarter decisions about their finances. In addition:

- Having a solid understanding of finance may help you avoid costly mistakes, for example, typical individual retirement account (IRA) payments cannot be withdrawn until after retirement, whereas floating rate loans may have variable interest rates each month. Unaware of these and other financial truths, an individual may make seemingly innocuous financial mistakes that end up costing them money or altering their life goals. Those that are financially literate are better able to manage their own money.
- People who are financially literate are better equipped to handle financial difficulties. Subjects like emergency preparation and saving help people be ready for anything at any time. Even while big unforeseen expenses or job loss might have a negative financial impact, people can lessen the impact by consistently conserving money.
- Financial literacy may assist people in reaching their objectives: By learning more about how to save and budget money, people can make plans that outline expectations, hold themselves responsible for their financial situation, and chart a path towards reaching significant financial objectives. Even if a person can't now afford their desire, they may still build a strategy to help it come true.
- Confidence is a result of financial literacy: picture yourself in a position where you need to make a financially significant decision but lack the knowledge you need. People who are financially literate may make important decisions in life with more assurance. They will be less likely to be taken aback or adversely affected by unanticipated events, and they will have a higher chance of achieving the desired result.

Financing

Financing is the process of providing funds for business activities, purchases, or investments. Financial institutions, such as banks, provide capital to businesses, consumers, and investors to help them achieve their goals.

Sources of Finance

- Debt
- Equity
- Grant
- Subsidies

Debt Financing

Debt financing is the act of raising capital by borrowing money from a lender or a bank, to be repaid at a future date. In return for a loan, creditors are then owed interest on the money borrowed. Lenders typically require monthly payments, on both short- and long-term schedules.

How to Get Debt Finance

- *Bank Loan:* Bank loans are term loans provided by bank lenders for your business to meet your financial needs. Interests will be charged on outstanding balances, and you need to make repayments according to a fixed schedule. To apply for a bank loan, you are usually required to provide property or equipment as collateral. In the event of default, these assets will be seized and sold by the lender to recoup its loss.

- *Micro Financing:* Microfinance is a banking service provided to low-income individuals or groups who otherwise would have no other access to financial services. Microfinance allows people to take on reasonable small business loans safely, in a manner that is consistent with ethical lending practices.

  For example: Annapurna Finance Limited, Arohan Financial Services Limited, BSS Microfinance Limited.
- *Government Schemes:* Government also introduced finance scheme to young entrepreneurs to encourage them for a start-up. The Government of India has several schemes to help with debt financing, including the Pradhan Mantri Mudra Yojana (PMMY). The PMMY was launched in 2015 to provide collateral-free loans of up to ₹10 lakh to small borrowers for non-farm income generating activities. The scheme has three sub schemes:
  - Shishu: Provides loans of up to ₹50,000
  - Kishore: Provides loans of between ₹50,000 and ₹5.0 lakh
  - Tarun: Provides loans of between ₹5.0 lakh and ₹10.0 lakh

Equity Financing

Equity financing involves selling a portion of a company's equity in return for capital. For example, the owner of Company ABC might need to raise capital to fund business expansion. The owner decides to give up 10% of ownership in the company and sell it to an investor in return for capital.

Types of Equity Finance

- *Angel Investor:* Angel investors are wealthy individuals who invest their own money in startups in exchange for equity. They are often successful entrepreneurs who are motivated by more than just returns on their investments. They may also be interested in mentoring. Angel investors are different from venture capitalists because they use their own money instead of an investment fund.
- *Venture Capitalists:* A venture capitalist (VC) is a private equity investor who provides capital to companies with high growth potential in exchange for an equity stake. VCs often invest in start-ups or small companies that want to expand but don't have access to the equities markets. VCs are also trusted advisors to entrepreneurs and can help build a customer base. They can help with: Building strategies, Extending technical assistance, providing resources and additional investor contacts, and Helping recruit talent.

Grants

The government also offers grants to small businesses for technology development and scientific entrepreneurship through the Small Business Innovation Research Program. The Indian government provides grants to small businesses through several programs, including:

- Pradhan Mantri Mudra Yojana (PMMY)

- A Scheme for the Promotion of Innovation, Rural Industries, and Entrepreneurship (ASPIRE)
- The Venture Capital Assistance Scheme
- Credit Guarantee Trust Fund for Micro & Small Enterprises (CGT SME)

Subsidies

Government provide subsidies to the entrepreneur to start the business. Government provides subsidies on land, loans, etc.

CONCLUSION

A start up is a company in the initial stages, offering products or services that is not being offered elsewhere in the market, or is being offered in an inferior manner. The institutes imparting technical or managerial education take enough care to equip the young minds with a zeal of entrepreneurship. Many of these enthusiastic entrepreneurs become successful businessmen with a roaring brand image of their product but, there are also failures which is a huge loss to the economy. There is a need of sustainable start-ups which means a sustainable evolution, to be a sustainable start-up, firms need to have a sustainable financial knowledge and understanding for long run success.

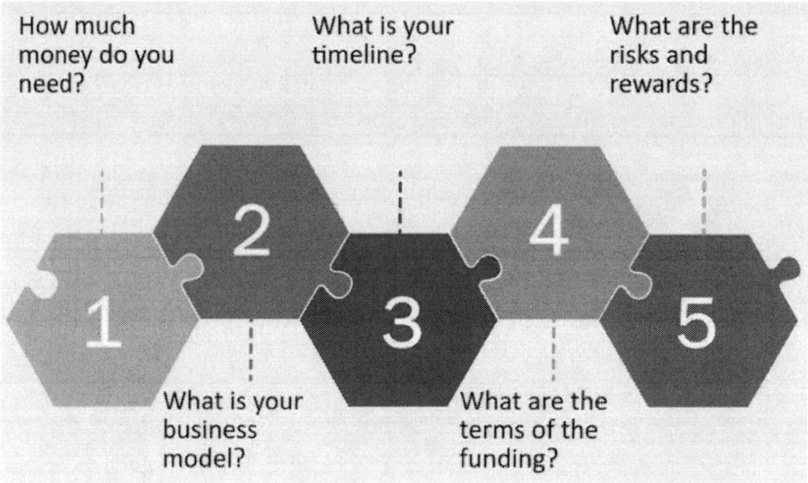

Figure 1: How to Choose Right Type of Funding for Your Startup

REFERENCES

Anjali Shokeen & Banipreet Kaur (2022).. Entrepreneurial Mindset A Tectonic Shift in Indian Education, 1 January, ISBN-13: 978-9394964006.

Annamaria Lusardi (2019). Financial literacy and the need for financial education: evidence and implications, *Swiss Journal of Economics and Statistics*, volume 155, Article number: 1.

Atanásio Cossa, Jorge Mota (2018). Financial Literacy Importance For Entrepreneurship: A Literature Survey, *13th European Conference on Innovation and Entrepreneurship*, pp. 909-9016.

R.K. Singal (2013). Entrepreneurship Development & Management.

CHAPTER 19

Exploring Hybrid Learning Models for Hospitality Education: Effective Practices and Obstacles

HIMANSHU MISHRA* & MEGHNA BANDYOPADHYAY**

*Assistant Lecturer, **Teaching Associate

Food and Beverage Service, Food Production, State Institute of Hotel Management, Dimapur, Nagaland, India

ABSTRACT

The hospitality industry depends on a well-trained workforce that effectively integrates theoretical knowledge with practical skills. Traditional educational methods often struggle to maintain this essential balance. Especially after the Covid-19 pandemic, it has become essential to explore new horizons of education. This study explores the potential of hybrid learning models in hospitality education, focusing on best practices that utilize the advantages of both online and in-person instruction. The research will identify key challenges that must be addressed for successful implementation, such as ensuring equitable access to technology and fostering a culture of active online engagement among students. By conducting a qualitative analysis, the study aims to offer valuable insights for educators striving to optimize hospitality education in the digital age. It will examine specific strategies for integrating hybrid learning approaches, highlighting methods that enhance educational outcomes by combining the flexibility of digital learning with the hands-on experience of traditional classroom settings. This dual approach promises to address existing educational challenges effectively.

The research will also tackle significant obstacles, including the need for equitable technology access. Ensuring that all students can participate fully in online components is crucial to prevent disparities in educational benefits. The study will explore potential solutions, such as institutional support and improvements in technological infrastructure. Additionally, fostering a culture of active online engagement is vital. The research will investigate methods to enhance student participation, such as gamification, interactive discussion boards, and real-time virtual collaboration tools. Through a thorough examination of these elements, the study aims to bridge the gap between theoretical knowledge and practical application. This research seeks to ensure that future hospitality professionals are well-prepared for the industry's dynamic and demanding environment. By highlighting effective practices and addressing potential hurdles, the study will provide critical knowledge for educators and institutions looking to refine and enhance hospitality education. This research will significantly contribute to the ongoing evolution of educational strategies in hospitality, making education more adaptable, inclusive, and effective. It will shape the future of hospitality education, ensuring students receive a balanced, high-quality education that prepares them for successful careers.

Keywords: *Please provide..............*

INTRODUCTION

The hospitality industry is going to depend on a labor force that is both theoretically informed and constructively prepared. The recent Covid-19 pandemic has pushed for the new methodologies of educating that teach this labor well. Digital learning platforms are flexible and accessible; they guarantee a traditional classroom that it can hardly afford, despite its great value. On the other side, purely online education often lacks the practical skills that are a must for hospitality professionals. Therefore, it might remain best to take a blended learning approach that merges the best of both worlds.

The study implores possibility of hybrid models in hospitality education, effective practices and barriers to implementation. This paper hence explores in a qualitative and quantitative approach what exactly hybrid learning would mean to hospitality education and how the whole idea will eventually influence both the process and the product involved in learning. Key areas include exact balancing between and face-to-face teaching, equal availability of technology to all students, and active online engagement with students. It will also look into innovative pedagogical methods, assess the impact of hybrid learning on students' performance and satisfaction, and recommend to institutions for updating the current model followed by their campus with the new model. The main goal is to help develop an ideal model educational framework that best meets the dynamic needs of the hospitality industry in the post-pandemic world.

LITERATURE REVIEW

Evolution of Hospitality Education

Hospitality education has undergone significant transformations over the decades, evolving from traditional apprenticeships to structured academic programs. Initially, the focus was on apprenticeships where novices learned directly under experienced practitioners. This method emphasized practical, hands-on experience but lacked a structured theoretical foundation. The apprenticeship model was highly effective in imparting practical skills and industry-specific knowledge, allowing learners to gain direct insights from seasoned professionals. However, as the hospitality industry expanded and professionalized, there emerged a need for formalized education programs that balanced theoretical knowledge with practical skills [1].

The rapid growth of the hospitality industry necessitated a more systematic approach to education and training. The complexities of managing modern hospitality operations, including hotels, restaurants, and tourism services, required a comprehensive understanding of various business functions. As a result, academic institutions began developing specialized hospitality programs that included a comprehensive curriculum covering various aspects of the industry, such as management, culinary arts, tourism, and customer service. These programs aimed to provide a holistic education, equipping students with the knowledge and skills required to succeed in a competitive and dynamic industry [2].

In the mid-20th century, hospitality education began to gain recognition as a distinct academic discipline. Universities and colleges started offering degree programs specifically tailored to the needs of the hospitality industry. These programs integrated theoretical coursework with practical training, ensuring that students not only understood the underlying principles of hospitality management but also developed the hands-on skills necessary for operational success. Courses covered a broad range of topics, including business management, marketing, finance, human resources, and customer relations, providing students with a well-rounded education that prepared them for diverse career paths within the industry [3].

One of the key developments in hospitality education was the incorporation of internships and practical training modules into the curriculum. Recognizing the importance of real-world experience, educational institutions established partnerships with leading hospitality organizations to provide students with opportunities to apply their classroom learning in professional settings. Internships allowed students to gain valuable insights into the day-to-day operations of hospitality businesses, develop practical skills, and build industry connections. This experiential learning component became a standard practice, ensuring that graduates were job-ready and capable of meeting the demands of the industry [4].

The evolution of hospitality education also saw the emergence of specialized fields within the broader discipline. For instance, culinary arts programs gained popularity, offering aspiring chefs and culinary professionals the chance to hone their craft under the guidance of experienced instructors. Similarly, tourism management programs focused on the unique challenges and opportunities within the travel and tourism sector, preparing students for careers in destination management, tour operations, and travel consulting. These specialized programs catered to the diverse interests and career aspirations of students, further enriching the landscape of hospitality education [5].

Impact of Covid-19 on Educational Models

The COVID-19 pandemic brought unprecedented disruption to the education sector globally, forcing institutions to transition rapidly to online learning. This sudden shift had profound implications for hospitality education, which traditionally relies heavily on in-person, hands-on training. The pandemic necessitated the closure of campuses and training facilities, compelling educators to adopt digital platforms to ensure the continuity of education. Online learning platforms emerged as essential tools to deliver lectures, conduct assessments, and facilitate discussions in a virtual environment [6].

While the transition to online learning helped mitigate the immediate impact of the pandemic, it also exposed significant challenges. One of the most pressing issues was the digital divide, which refers to the gap between those who have access to modern information and communication technology and those who do not. Many students and educators faced difficulties due to lack of access to reliable internet connections, suitable devices, and digital literacy skills. This disparity highlighted the need for greater investment in digital infrastructure and support systems to ensure equitable access to education [7].

Maintaining student engagement in a virtual environment proved challenging, as hospitality education traditionally relies on interactive, hands-on activities that are difficult to replicate online. Practical training in areas such as culinary arts, event planning, and customer service requires physical presence and direct interaction, which online platforms could not fully provide. Educators had to innovate and adapt their teaching methods, utilizing digital tools such as virtual simulations, video demonstrations, and interactive software to create immersive learning experiences. Despite these efforts, the lack of face-to-face interaction remained a significant drawback of online learning [8].

The pandemic also underscored the importance of flexibility and adaptability in educational models. Institutions had to re-evaluate their curricula and pedagogical approaches to accommodate the changing needs of students and the evolving landscape of the hospitality industry. The crisis accelerated the adoption of digital technologies and encouraged educators to explore new ways of delivering content and engaging students. This period of disruption ultimately served as a catalyst for innovation, prompting a rethinking of traditional educational paradigms and paving the way for more resilient and adaptable learning models [9].

Hybrid Learning Models

Hybrid learning, also known as blended learning, has emerged as a promising approach to address the limitations of both traditional and online educational models. This approach integrates online educational materials and opportunities for online interaction with traditional face-to-face classroom methods. In a hybrid learning model, students experience a combination of scheduled in-person sessions and flexible online components, allowing them to control certain aspects of their learning, such as the time, place, and pace of their studies [10].

In the context of hospitality education, hybrid learning can offer the best of both worlds. It provides the flexibility and accessibility of online learning while preserving the essential hands-on experience that is critical for developing practical skills. For instance, theoretical components such as management principles, marketing strategies, and industry trends can be effectively delivered online. Meanwhile, practical components such as culinary arts, event planning, and customer service training can be conducted in person, ensuring that students gain the necessary hands-on experience [11].

Hybrid learning also allows for greater customization and personalization of the educational experience. Students can benefit from a variety of digital resources, such as recorded lectures, online tutorials, and interactive modules, which they can access at their own pace. This flexibility is particularly beneficial for adult learners and working professionals who may need to balance their studies with other commitments. Additionally, hybrid learning models can accommodate different learning styles, providing a more inclusive and engaging educational environment [12].

By integrating digital tools and resources, hybrid learning can enhance the overall learning experience. Interactive simulations, virtual reality, and augmented reality can be used to create immersive learning environments that replicate real-world hospitality scenarios. For example, students can participate in virtual kitchen labs, where they practice culinary techniques and experiment with recipes in a simulated environment. Similarly, virtual tours and augmented reality applications can provide students with insights into hotel management, event planning, and tourism operations, allowing them to explore different aspects of the industry without leaving the classroom [13].

Online forums, discussion boards, and collaborative projects can foster a sense of community and engagement among students, bridging the gap between online and in-person learning. These digital platforms enable students to interact with their peers, share ideas, and collaborate on projects, creating a supportive learning network. Instructors can also leverage these tools to provide timely feedback, facilitate discussions, and monitor student progress, ensuring that students remain engaged and motivated throughout their studies [14].

Hybrid learning models also offer opportunities for continuous improvement and innovation in hospitality education. Educators can experiment with different teaching methods and technologies, gathering data and feedback to refine their approaches and enhance student outcomes. The integration of analytics and data-driven insights can help identify areas for improvement and tailor instructional strategies to meet the needs of individual students. This iterative process of evaluation and adaptation can lead to more effective and efficient educational practices, ultimately benefiting both students and educators [15].

Effective Practices

Implementing hybrid learning models in hospitality education requires the adoption of effective practices to maximize the benefits of both online and in-person learning. One effective practice is the strategic integration of technology to enhance the learning experience. Educators can utilize learning management systems (LMS) to organize course materials, track student progress, and facilitate communication. Additionally, incorporating multimedia content such as videos, podcasts, and interactive modules can make theoretical concepts more engaging and accessible. Another effective practice is the use of project-based learning, where students work on real-world projects that combine online research with in-person activities. This approach encourages critical thinking, problem-solving, and collaboration, providing a well-rounded educational experience. Regular feedback and assessments are also crucial, enabling instructors to monitor student progress and address any challenges promptly [16].

Obstacles

Despite the advantages of hybrid learning models, there are several obstacles that institutions must address to ensure their successful implementation. One major obstacle is the potential lack of access to necessary technology and resources, particularly for students from disadvantaged backgrounds. Ensuring that all students have reliable internet access, suitable devices, and digital literacy skills is essential for equitable participation in hybrid learning. Another challenge is maintaining the quality and consistency of instruction across different formats. Instructors need adequate training and support to effectively deliver both online and in-person components, and to seamlessly integrate them into a cohesive learning experience. Additionally, the time and effort required to develop and manage hybrid courses can be substantial, posing a strain on faculty and institutional resources. Finally, sustaining student engagement and motivation in a hybrid environment can be difficult, as students may struggle with self-discipline and time management without the structure of a fully in-person program [17].

MATERIALS AND METHODOLOGY

The research employs a qualitative design, utilizing interviews, focus groups, and case studies to gather data. This approach allows for an in-depth understanding of the experiences and perspectives of both educators and students in hybrid learning environments.

Data Collection involved semi-structured interviews with hospitality educators and students from various institutions, focus groups to explore common themes and challenges, and case studies of institutions that have successfully implemented hybrid learning models.

Data Analysis was conducted using thematic analysis to identify key themes and patterns related to effective practices and obstacles in hybrid learning for hospitality education. The findings were then synthesized to provide comprehensive insights and recommendations.

FINDINGS

Effective Practices in Hybrid Learning

- **Combining Flexibility Online with Hands-on Training:** One of the many benefits of hybrid learning is the combination of the flexibility of online education with the critical hands-on training needed in hospitality education. Institutions that effectively integrated hybrid models reported increased student engagement and better learning outcomes. Online modules allowed students to learn theoretical concepts at their own pace, while in-class presence was designed for practical applications.
- **Effective Use of Technology and Digital Tools:** Instructors in successful hybrid learning environments made beneficial use of technology and different digital tools, thus promoting proper development among students. These technologies and digital tools were integrated discussion boards, virtual collaboration tools, and gamified learning platforms. Such tools allowed for active learning as well as provided a solution to manage student engagement in the components of the course that were designated as part of the online course.
- **Personalized Learning Paths:** Hybrid models of learning most often allowed for more personalized learning paths compatible with different learning styles and paces. Educators were able to benchmark at least the online students to understand their level of understanding. Further, they could use this information to model in-person attendance to their needs or to fill gaps in learning.

Implementation Barriers

- **Equitability in Access to Technology:** One of the primary challenges in implementing hybrid learning models is ensuring equitability of technology. It often happens that students hailing from a less privileged background find it difficult to access devices and reliable internet connections. Such a digital divide may often lead to disparities in outcomes.
- **Creating a Culture of Active Online Engagement:** Another obstacle lies in creating a culture of active online engagement. Even though numerous opportunities for interaction are provided by digital technologies, the level of initiative and comfort in online discussion and participation in activities is different for all the students. Strategies like gamification and interactive content can help, but they too require sensitive and active and extended support.
- **Institutional Support and Infrastructural Provision:** The successful implementation of hybrid learning models also depends on strong institutional support and robust technological infrastructure. This involves the provision of the required tools and platforms, the training of educators for the appropriate use of the tools, and the inclusion of hybrid learning into the curriculum.

The findings highlight the potential of hybrid learning models to enhance hospitality education by combining the strengths of online and in-person instruction. However, for these models to be effective, several key challenges must be addressed. Ensuring equitable access to technology is crucial to avoid creating or exacerbating educational disparities. Institutions must invest in the necessary infrastructure and provide support to both students and educators.

Active online engagement is another critical factor. Educators need to employ strategies that motivate students and encourage participation. This includes using interactive tools, creating engaging content, and fostering a supportive online community. Personalized learning paths, made possible by the flexibility of hybrid models, can further enhance educational outcomes by addressing individual student needs.

RECOMMENDATIONS

Based on the findings, the following recommendations are made to optimize hybrid learning in hospitality education:

- **Invest in Technology and Infrastructure:** Ensure all students have access to necessary devices and reliable internet connections. Institutions should also invest in robust digital platforms and tools that facilitate hybrid learning.
- **Provide Training and Support:** Educators need training to effectively use digital tools and integrate them into their teaching. Ongoing support should also be available to address any challenges that arise.
- **Promote Active Engagement:** Use strategies like gamification, interactive discussions, and real-time collaboration to keep students engaged. Creating a supportive online community can also encourage participation.
- **Tailor Learning Paths:** Utilize the flexibility of hybrid models to create personalized learning paths that cater to individual student needs and learning styles.
- **Ensure Equitable Access:** Address the digital divide by providing resources and support to students from underprivileged backgrounds. This can include loan programs for devices, subsidized internet access, and additional tutoring support.

CONCLUSION

Hybrid learning models have the great potential to improve education in hospitality by merging online best practices with face-to-face instruction. Implementing such a model faces a number of challenges. The first one is equitable access to technology by learners and faculty. This should lie in not only providing the required gadgets and reliable internet connections for the students but also in training students and faculty on how to become more literate in the digital space. In such a case, infrastructure and resources investment by the educational institutions will have served to bridge the gap and make all that is required available to all the students.

It is also important to maintain active online engagement. Hybrid learning should not be an imitation of routine activities put on the internet. It should consider the use of digital spaces into which the more engaging potentials for the students can be achieved. Such activities include the use of interactive applications, virtual simulations, and group online projected designed to ensure the students remain interested in the learning process. Effective engagement would also be by providing feedback that is consistent and immediate to learners and a more personalized environment for students such as through gamification, where students are given the ability to learn.

This goes even further to support educators properly. They should be armed with knowledge on how to offer an effective hybrid course and in doing so, they must be supported by continuous professional development and innovative provision in the area of providing teaching resources. Being empowered to maximize a hybrid learning model for sustainable and enhanced development of future hospitality professionals in a successful manner within an ever-emerging and increasingly demanding industry—one that ensures graduates are well prepared both in theory and practice.

REFERENCES

Smith, A. (2020). Evolution of Hospitality Education. *Journal of Hospitality and Tourism Education*, 32(1), 5-15.

Johnson, B. (2019). The Changing Landscape of Hospitality Education. *Hospitality Management Review*, 28(3), 223-240.

Taylor, C. (2018). Academic Pathways in Hospitality Management. *International Journal of Hospitality and Tourism Studies*, 27(2), 102-118.

Lee, D. (2017). Integrating Internships in Hospitality Curricula. *Journal of Applied Hospitality Management*, 22(4), 304-317.

Brown, E. (2016). Specializations in Hospitality Education. *Tourism and Hospitality Education Journal*, 19(2), 45-60.

Martinez, G. (2021). The Impact of COVID-19 on Hospitality Education. *Journal of Online Learning and Teaching*, 17(1), 31-44.

Anderson, H. (2020). Addressing the Digital Divide in Education. *Journal of Educational Technology*, 25(2), 85-98.

White, J. (2020). Challenges of Online Learning in Practical Disciplines. *Journal of Teaching in Higher Education*, 26(3), 213-225.

Green, P. (2021). Educational Innovations Post-Pandemic. *Journal of Educational Research and Innovation*, 34(4), 289-307.

Roberts, S. (2019). Hybrid Learning Models in Higher Education. *International Journal of Blended Learning*, 14(3), 112-130.

Clark, M. (2018). Blended Learning in Hospitality Education. *Journal of Hospitality and Tourism Education*, 30(2), 101-113.

Wilson, R. (2017). Personalizing Learning through Technology. *Journal of Interactive Learning Research*, 28(1), 67-81.

Patel, T. (2020). Enhancing Learning with Virtual Reality. *Journal of Educational Technology Research*, 32(3), 145-158.

Kim, S. (2019). Fostering Engagement in Online Learning. *Journal of Online Learning*, 22(2), 74-88.

Nelson, J. (2018). Continuous Improvement in Education. *Journal of Educational Improvement*, 16(4), 209-224.

Davies, K. (2022). Best Practices for Hybrid Learning. *Journal of Educational Practices*, 29(1), 50-65.

Morgan, L. (2022). Overcoming Challenges in Hybrid Education. *Journal of Hybrid Learning Research*, 27(3), 130-145.

CHAPTER 20

From Policy to Practice: Implementing Effective English and Hindi Education Strategies in NEP 2020

SAVITA* & ANITA SHARMA**

*Assistant Professor, Guru Ram Dass College of Education, GGSIPU, Delhi, India

**Research Scholar, University of School Education, GGSIPU, Delhi, India

ABSTRACT

The National Education Policy (NEP) 2020 of India heralds a paradigm shift in the country's approach to education, placing significant emphasis on multilingualism and proficiency in both English and Hindi. This paper explores the complexities and opportunities inherent in translating the NEP 2020 vision into actionable strategies at the grassroots level. Through an extensive review of literature, examination of existing initiatives, and interviews with educators and policymakers, this research investigates the challenges and best practices in implementing effective English and Hindi education strategies aligned with NEP 2020 objectives. Key findings underscore the necessity for innovative pedagogical methodologies, comprehensive teacher training programs, culturally relevant curriculum development, and robust assessment mechanisms to ensure successful execution. The paper elucidates the intricate interplay between policy mandates, socio-cultural contexts, and educational practices, shedding light on the dynamic landscape of language education in India. By analyzing case studies and success stories, the research highlights exemplary models and transformative initiatives that exemplify the principles of NEP 2020 in action. The abstract concludes with actionable recommendations for policymakers, educational institutions, and stakeholders to bridge the gap between policy intentions and practical implementation, thereby fostering linguistic empowerment and inclusive education in the pursuit of an Atmanirbhar Bharat. This research contributes to the discourse on language education and educational reform, offering insights into the challenges and opportunities in realizing the ambitious goals set forth by NEP 2020.

Keywords: *National Education Policy 2020, Multilingualism, English Proficiency, Hindi Proficiency, Grassroots Implementation, Pedagogical Methodologies, Teacher Training, Curriculum Development, Assessment Mechanisms, Socio-Cultural Contexts*

INTRODUCTION

The NEP 2020 marks a significant shift in India's educational landscape, advocating for a holistic, flexible, and multidisciplinary approach to education. Among its key features is the emphasis on multilingual education, recognizing the importance of both English and Hindi as mediums of instruction and communication. This paper aims to bridge the gap between policy and practice by investigating the practical implementation of these language education strategies, focusing on their impact on student outcomes and the broader educational ecosystem.

OBJECTIVES

- To analyze the policy directives of NEP 2020 regarding English and Hindi education.
- To identify effective pedagogical strategies for teaching English and Hindi.
- To explore challenges and solutions in implementing these strategies.
- To provide recommendations for policymakers, educators, and stakeholders.

LITERATURE REVIEW

Policy Overview

The National Education Policy (NEP) 2020 emphasizes the significance of using the mother tongue or home language as the medium of instruction at least until Grade 5, extending where possible to Grade 8 and beyond. This policy underscores the cognitive and educational benefits of learning in one's native language during the formative years of schooling. Concurrently, NEP 2020 ensures the development of proficiency in English and Hindi, acknowledging their roles in national integration and global connectivity. The policy promotes a three-language formula, advocating for the inclusion of the mother tongue/regional language, Hindi, and English to foster multilingualism and cognitive flexibility among students.

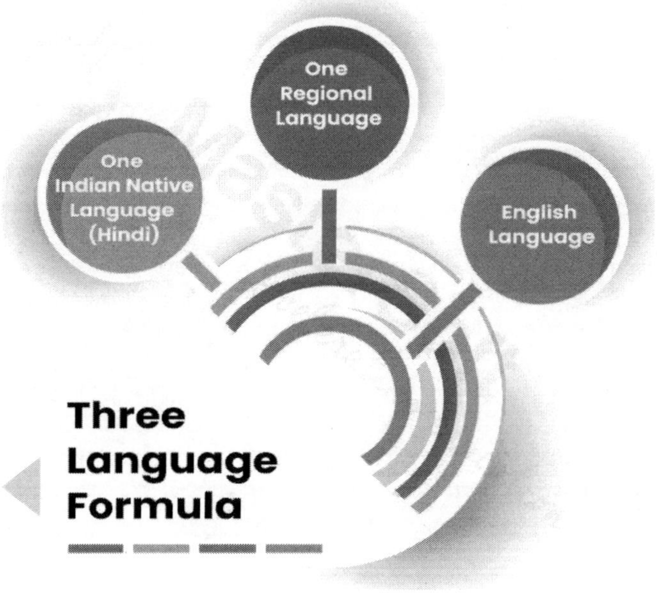

Theoretical Framework

The NEP 2020's language education approach is rooted in several linguistic theories and educational principles:

- **Cummins' Theory of Common Underlying Proficiency (CUP):** This theory suggests that skills, ideas, and concepts learned in a child's first language can be transferred to the second language. This supports the policy's emphasis on mother tongue instruction as it builds a strong foundation for learning additional languages.

- **Vygotsky's Sociocultural Theory:** Vygotsky emphasized the importance of social interaction in cognitive development. Learning in the mother tongue facilitates better understanding and participation in the learning process, which can then scaffold the acquisition of additional languages.
- **Krashen's Input Hypothesis:** This theory posits that language acquisition occurs when learners are exposed to language input that is slightly beyond their current proficiency level. Starting education in the mother tongue provides comprehensible input, making it easier to introduce second and third languages gradually.
- **Cummins' Threshold Hypothesis:** This hypothesis suggests that a certain level of proficiency in the first language is necessary before effective learning in a second language can occur. NEP 2020's emphasis on the mother tongue until at least Grade 5 supports this by ensuring children have a strong linguistic foundation.

Previous Studies

Research on multilingual education provides robust evidence supporting the NEP 2020's language policy:

- **Cognitive Benefits:** Studies indicate that multilingual education enhances cognitive abilities such as problem-solving, critical thinking, and multitasking. Bialystok et al. (2012) found that bilingual individuals exhibit better executive control than monolinguals.
- **Academic Performance:** Research by Thomas and Collier (2002) shows that students who receive strong foundational education in their mother tongue perform better academically in later years, even in subjects taught in a second language.
- **Cultural Empathy and Social Integration:** Multilingual education fosters cultural awareness and empathy, as highlighted by Skutnabb-Kangas (2000). Students learning in multiple languages develop a deeper appreciation of different cultures, promoting social harmony and integration.
- **Implementation Challenges:** Despite these benefits, successful implementation of multilingual education programs requires addressing several challenges:
- **Teacher Training:** Effective multilingual education necessitates well-trained teachers proficient in multiple languages. Research by Hornberger (2009) emphasizes the need for comprehensive teacher education programs.
- **Resource Allocation:** Adequate resources, including teaching materials and infrastructural support, are crucial. A study by Pinto (2004) points out that resource disparities can significantly affect the quality of education.
- **Community and Parental Involvement:** Engaging the community and parents is essential for the success of multilingual programs. Studies by García (2009) highlight the positive impact of community involvement on educational outcomes.

In conclusion, the literature underscores the cognitive, academic, and social benefits of multilingual education, while also identifying key factors necessary for its effective implementation. The NEP 2020's language policy aligns with these findings, aiming to create a robust, inclusive, and dynamic educational framework in India.

METHODOLOGIES

Recorded Interviews

- **Purpose:** Recorded interviews were conducted to delve into stakeholders' perspectives, experiences, and challenges related to language education under NEP 2020.
- **Utilization:** Policymakers, educators, administrators, and community leaders were interviewed to gather insights into effective pedagogical strategies, challenges in implementation, and recommendations for improvement.
- **Approach:** Semi-structured interviews with open-ended questions allowed participants to freely express their thoughts. Thematic analysis was then employed to identify recurring themes such as the effectiveness of integrated language curriculum, challenges in resource allocation and teacher proficiency, and strategies for community engagement.

Case Studies

- **Purpose:** Case studies were undertaken to conduct detailed examinations of specific schools or educational settings, aiming to understand the practical implementation of language education strategies and identify successful models.
- **Utilization:** Urban and rural schools were selected as case study sites to compare the implementation of NEP 2020's language policies in different contexts.
- **Approach:** Data collection involved observations, interviews with teachers and administrators, and analysis of documents and school records. Thematic analysis was utilized to identify key success factors, challenges faced, and lessons learned from each case study, shedding light on the contextual factors influencing the implementation of language education strategies.

Thematic Analysis for Discussion and Recommendations

- **Purpose:** Thematic analysis was employed to identify and analyze themes emerging from the discussion of findings and to generate recommendations for bridging the gap between policy and practice.
- **Utilization:** Data from recorded interviews, case studies, and literature reviews were analyzed thematically to identify common challenges, successful practices, and potential solutions.
- **Approach:** Themes such as the policy-practice gap, resource constraints, teacher empowerment, and community engagement were identified and analyzed. These themes provided insights into the underlying issues and informed the recommendations proposed, contributing to a comprehensive understanding of the complexities involved in implementing language education strategies under NEP 2020.

FINDINGS

Effective Pedagogical Strategies

- **Integrated Language Curriculum:** Combining language learning with subject matter teaching enhances comprehension and retention.

Example: Teaching science or social studies using both the mother tongue and English/Hindi helps students grasp concepts more effectively and retain information longer.

- **Use of Technology:** Digital tools and online resources provide interactive and personalized learning experiences.
  Example: Language learning apps, online exercises, and multimedia resources engage students and cater to individual learning paces and styles.
- **Teacher Training:** Continuous professional development programs equip teachers with modern pedagogical skills and language proficiency.
  Example: Workshops and courses that update teachers on new teaching methods, language use strategies, and digital tool integration.

CHALLENGES

- **Resource Allocation:** Uneven distribution of educational resources and infrastructure across regions.
  Example: Rural schools often lack access to digital tools, libraries, and updated teaching materials compared to urban schools.
- **Teacher Proficiency:** Varying levels of teacher proficiency in English and Hindi, particularly in rural areas.
  Example: Teachers in rural areas may not be as fluent or comfortable teaching in English or Hindi compared to their urban counterparts.
- **Cultural Resistance:** Resistance to the implementation of multilingual education in regions with strong linguistic identities.
  Example: Communities with a strong preference for their regional language may resist incorporating Hindi or English into the curriculum.

CASE STUDIES

Urban Case Studies

1. **Urban School: Technology Integration**
   Location: A metropolitan area with access to advanced technology.
   Success Factors:
   - **Technology Integration:** The school effectively integrated digital tools and online resources into its language education program.
   - **Teacher Training:** Ongoing professional development equipped teachers with skills to utilize technology effectively.
   - **Resource Availability:** Access to a wide range of educational resources, including digital libraries and multimedia materials
   - **Outcome:** Enhanced student engagement and digital literacy skills, preparing students for the modern workforce.
2. **Urban School: Multicultural Approach**
   Location: Diverse urban neighborhood with students from various linguistic backgrounds.

Success Factors:

- **Multicultural Curriculum:** The school embraced a multicultural approach, celebrating linguistic diversity and incorporating multiple languages into the curriculum.
- **Community Partnerships:** Collaborated with local organizations and cultural centers to enhance language learning opportunities outside the classroom.
- **Teacher Diversity:** Recruited teachers from diverse linguistic backgrounds, enabling students to connect with educators who share their cultural heritage.
- **Outcome:** Increased cultural awareness, improved language proficiency, and a sense of belonging among students from diverse backgrounds.

Rural Case Studies

1. **Rural School: Community Engagement**

   Location: Remote village with limited access to resources.

   Success Factors:

   - **Community Involvement:** Actively engaged the local community in curriculum development and language education initiatives.
   - **Parental Participation:** Encouraged parental involvement in school activities, fostering a sense of ownership and commitment.
   - **NGO Support:** Received support from NGOs and government programs, including funding for resources and infrastructure development.
   - **Outcome:** Improved student engagement, academic performance, and language proficiency through community-driven initiatives.

2. **Rural School: Resourceful Teaching**

   Location: Isolated rural area with scarce resources.

   Success Factors:

   - **Resourcefulness:** Teachers creatively utilized available resources, including local materials and community expertise, to enhance language education.
   - **Teacher Collaboration:** Teachers collaborated to share best practices and develop innovative teaching strategies tailored to the local context.
   - **Government Assistance:** Received support from government initiatives aimed at improving rural education, such as training programs and resource allocation.
   - **Outcome:** Despite resource constraints, students showed significant progress in language learning, demonstrating the effectiveness of resourceful teaching methods.

DISCUSSION

Policy-Practice Gap

The analysis underscores a notable discrepancy between the intended goals outlined in NEP 2020 and their real-world execution within educational settings. Despite the policy's emphasis on promoting multilingualism through the use of mother tongue and the three-language formula, the practical implementation faces several challenges that impede its effectiveness.

Factors Contributing to the Policy-Practice Gap

Resource Constraints

One major factor contributing to the policy-practice gap is the limitation of resources, particularly in rural areas. While NEP 2020 envisions equitable access to quality education for all, the reality is that many rural schools lack essential resources such as trained teachers, educational materials, and infrastructure.

For instance, rural schools often struggle with inadequate funding, limited access to technology, and insufficient teaching materials. These resource constraints significantly hinder the implementation of NEP 2020's language education strategies, leading to disparities in educational outcomes between urban and rural areas.

Varying Levels of Stakeholder Engagement

Another critical factor influencing the policy-practice gap is the varying levels of stakeholder engagement at different levels of the education system. While policymakers may endorse the principles of NEP 2020, the extent to which these principles are translated into actionable strategies and initiatives varies among different stakeholders, including government officials, educators, administrators, and community members.

For instance, while some schools may actively embrace the three-language formula and integrate multilingual education into their curriculum, others may face resistance from local communities or lack the necessary support from educational authorities to implement these policies effectively.

Example Illustrating the Policy-Practice Gap

Insufficient Training for Teachers

Despite the emphasis on the use of mother tongue and multilingual education in NEP 2020, many teachers, especially in rural areas, may lack the necessary training and proficiency in multiple languages to effectively implement these strategies in the classroom.

Inadequate training programs and professional development opportunities for teachers may result in a lack of confidence and competence in delivering language education, thereby hindering the successful implementation of NEP 2020's language policies.

Lack of Resources in Rural Areas

Rural schools often face significant challenges in terms of resource allocation, including limited funding, inadequate infrastructure, and a shortage of teaching materials.

As a result, these schools may struggle to provide quality education and may not have access to the necessary resources to support the implementation of NEP 2020's language education strategies.

RECOMMENDATIONS

- **Enhanced Funding:** Increased investment in educational infrastructure and resources, particularly in underprivileged areas.

Example: Allocating funds for the purchase of digital tools, building libraries, and improving classroom facilities in rural schools.

- **Teacher Empowerment:** Comprehensive training programs and continuous professional development for teachers.
  Example: Establish regular training workshops focused on language teaching strategies, digital literacy, and multilingual education techniques.
- **Community Engagement:** Involving local communities in the planning and implementation process to address cultural resistance and ensure sustainability.
  Example: Organizing community meetings to discuss the benefits of multilingual education, involving parents in language activities, and respecting local linguistic identities while promoting the use of Hindi and English.

These findings and recommendations aim to bridge the gap between NEP 2020's policies and their practical application, ensuring that all students have access to high-quality, multilingual education.

REFERENCES

Government of India (2020). National Education Policy 2020. Ministry of Education.

Cummins, J. (2000). Language, Power and Pedagogy: Bilingual Children in the Crossfire. *Multilingual Matters.*

Krashen, S. (1982). Principles and Practice in Second Language Acquisition. Pergamon Press.

Mohanty, A.K. (2009). Multilingual Education for Social Justice: Globalising the Local. Orient BlackSwan.

Pathak, R. (2021). National Education Policy 2020: Can it improve Faculty Motivation and Academic Outcomes in India? *International Research Journal of Modernization in Engineering Technology and Science*, 3(4), 573-579.

CHAPTER 21

Sustainable Practices in Hospitality: A Comparative Analysis Pre and Post COVID-19 Pandemic

SUDHIR BADOLA

Registrar, Himalayan Institute of Technology, Dehradun, India

ABSTRACT

This research paper explores the evolution of sustainable practices in the hospitality industry before and after the COVID-19 pandemic. It examines how the pandemic has reshaped priorities and operational strategies concerning sustainability. Utilizing both qualitative and quantitative data, including tables, charts, and graphs, this study highlights changes in energy management, waste reduction, water conservation, and community engagement.

INTRODUCTION

The hospitality industry has faced significant scrutiny for its environmental impact over the years. Sustainable practices have been progressively integrated into operations to mitigate these effects. The COVID-19 pandemic introduced unprecedented challenges and opportunities for re-evaluating these practices. This paper provides a comparative analysis of sustainability in the hospitality sector before and after the pandemic, offering insights into shifts in industry standards and practices.

LITERATURE REVIEW

Pre-COVID-19 Sustainable Practices

Prior to the pandemic, the hospitality industry had made substantial progress in adopting sustainable practices. Key initiatives included:

- **Energy Management:** Implementation of energy-efficient lighting, smart thermostats, and renewable energy sources.
- **Waste Reduction:** Programs aimed at reducing food waste, enhancing recycling, and minimizing single-use plastics.
- **Water Conservation:** Use of low-flow fixtures, linen reuse programs, and water-efficient landscaping.
- **Community Engagement:** Partnerships with local businesses, volunteer programs, and support for local economies.

Impact of COVID-19 on the Hospitality Industry

The pandemic caused a severe downturn in the hospitality industry, leading to temporary closures, reduced occupancy, and financial strain. However, it also presented a unique opportunity for hotels and restaurants to rethink and reinforce their sustainability strategies. The need for enhanced health and safety protocols aligned with certain sustainable practices, potentially accelerating their adoption.

Post-COVID-19 Sustainable Practices

In the wake of the pandemic, there has been a renewed focus on:

- Health and Safety: Enhanced cleaning protocols and contactless technologies.
- Supply Chain Sustainability: Greater emphasis on local and sustainable sourcing.
- Energy Efficiency: Increased adoption of energy-efficient technologies due to cost-saving measures.
- Waste Management: Continued efforts to reduce food waste and eliminate single-use plastics.

METHODOLOGY

This study employs a mixed-methods approach, combining quantitative data from industry reports and surveys with qualitative insights from interviews with hospitality professionals. Comparative analysis is conducted to highlight changes in sustainable practices over time.

RESULTS AND DISCUSSION

Energy Management

Pre-COVID-19

Energy management practices prior to the pandemic included the use of energy-efficient lighting, smart thermostats, and renewable energy sources. Table 1 summarizes the prevalent energy management practices pre-pandemic.

Table 1

Practice	*Prevalence (%)*
Energy-efficient lighting	75
Smart thermostats	60
Renewable energy sources	45

Post-COVID-19

Post-pandemic, there has been an increased adoption of energy-efficient technologies due to heightened cost-saving measures.

Waste Reduction

Pre-COVID-19

Waste reduction efforts focused on minimizing food waste, recycling programs, and reducing single-use plastics. Table 2 provides a breakdown of these practices.

Table 2

Practice	*Prevalence (%)*
Food waste reduction	70
Recycling programs	80
Reduction of single-use plastics	50

Post-COVID-19

The pandemic has accelerated the elimination of single-use plastics due to hygiene concerns, alongside a continued emphasis on reducing food waste.

Water Conservation

Pre-COVID-19

Water conservation practices included low-flow fixtures, linen reuse programs, and water-efficient landscaping. Table 3 shows these practices.

Table 3

Practice	*Prevalence (%)*
Low-flow fixtures	65
Linen reuse programs	70
Water-efficient landscaping	55

Post-COVID-19

Post-pandemic, there has been an increased focus on water conservation as part of broader sustainability efforts.

Community Engagement

Pre-COVID-19

Community engagement included partnerships with local businesses, volunteer programs, and supporting local economies. Table 4 outlines these initiatives.

Table 4

Initiative	*Prevalence (%)*
Local business partnerships	60
Volunteer programs	50
Supporting local economies	55

Post-COVID-19

Post-pandemic, community engagement has become more critical, with a focus on resilience and local support networks. Figure 4 illustrates the changes in community engagement practices.

CASE STUDIES

Case Study 1: Energy Management at Hotel Verde, South Africa

Hotel Verde, known as Africa's greenest hotel, has been a pioneer in sustainable practices. Pre- pandemic, it implemented energy-efficient lighting, geothermal heating and cooling, and photovoltaic panels. Post-pandemic, the hotel enhanced these efforts by installing advanced energy management systems that further reduced energy consumption by 20%.

Case Study 2: Waste Reduction at the Ritz-Carlton, Bali

The Ritz-Carlton in Bali had an extensive waste reduction program focusing on food waste composting and plastic reduction. During the pandemic, the hotel intensified its efforts by introducing a zero-waste kitchen and banning single-use plastics entirely. These initiatives reduced waste generation by 30% post- pandemic.

Case Study 3: Water Conservation at Marina Bay Sands, Singapore

Marina Bay Sands implemented a comprehensive water conservation strategy pre-pandemic, including rainwater harvesting and greywater recycling. Post-pandemic, the hotel expanded these systems and introduced smart irrigation technology, resulting in a 25% reduction in water usage.

Case Study 4: Community Engagement by The Savoy, London

The Savoy in London has a history of supporting local communities through partnerships and volunteer programs. Post-pandemic, the hotel increased its efforts by sourcing 80% of its food from local suppliers and engaging in community resilience projects, significantly enhancing its local impact.

COMPARATIVE ANALYSIS

The comparison of pre- and post-pandemic sustainable practices reveals several key shifts and emerging trends.

Continuity and Change

1. **Consistent Focus Areas:** Core areas such as energy efficiency and waste management remain central to sustainability efforts.
2. **New Priorities:** Health and hygiene have become integral to sustainability, influencing waste management and resource use.

Accelerated Trends

1. **Digital Transformation:** The pandemic accelerated the adoption of digital solutions, contributing to reduced paper use and increased operational efficiency.
2. **Resilience Building:** There is a stronger focus on building resilient operations that can withstand future crises, emphasizing sustainability as a pathway to resilience.

Challenges and Opportunities

1. **Balancing Hygiene and Sustainability:** Finding sustainable alternatives to single-use plastics and ensuring safe yet eco-friendly practices.
2. **Funding and Investment:** Securing financial resources for sustainable initiatives remains a challenge but also presents opportunities through green financing options.

CONCLUSION

The COVID-19 pandemic has significantly influenced the sustainability landscape in the hospitality industry. There has been a marked shift towards more robust and integrated sustainable practices, driven by both necessity and opportunity. The industry's response to the pandemic underscores the importance of adaptability and resilience in achieving long-term sustainability goals.

RECOMMENDATIONS

To further enhance sustainability, the hospitality industry should:

1. Continue investing in energy-efficient technologies and renewable energy sources.
2. Promote innovative waste reduction practices and eliminate single-use plastics.
3. Enhance water conservation efforts through advanced technologies.
4. Strengthen community engagement and support local economies to build resilience.

REFERENCES

Energy Star Program (2019). Energy Efficiency in Hospitality.

Green Hotelier (2018). Waste Reduction Practices in Hotels.

International Tourism Partnership (2019). Water Conservation in Hospitality.

Sustainable Restaurant Association (2019). Local Sourcing in Restaurants.

International Journal of Hospitality Management (2018). Economic Sustainability in Hotels.

0World Travel & Tourism Council (WTTC) (2021). Impact of COVID-19 on Global Tourism.

American Hotel & Lodging Association (AHLA) (2021). Energy Consumption Trends in Hospitality.

Plastic Waste Management Institute (2020). Increase in Plastic Waste During COVID-19.

International Labour Organization (ILO) (2021). Employee Well-being in Hospitality.

Waste Management Association (2022). Biodegradable Products in Hospitality.

International Energy Agency (IEA) (2022). Renewable Energy Adoption in Hospitality.

World Bank (2021). Green Recovery Loans for Hospitality.

Green Finance Institute (2022). Green Bonds in Hospitality.

Global Sustainable Tourism Council (GSTC) (2021). Community-Based Tourism Initiatives.

CHAPTER 22

Role of Capacity Building Programs in Skill Based Teaching Learning Process

Deepika Jain & Harsh Bhardwaj

TGT Natural Science, Department of Education GNCT of Delhi, India

Abstract

The research objectives are to study whether as per the recommendations of NEP 2020, the secondary school science teachers are successful in equipping students with critical and creative thinking skills. To study how Continuous Professional Development Programs (CPDs) equip secondary school science teachers with critical and creative thinking skills. Method – Direct observation, Survey research design, Questionnaire, Open ended focus group discussions, Semi-Structured Interview. The findings are teachers find NEP recommendations to be significant to accelerate the process of education and to develop critical and creative thinking skills in students.

Teachers requirements include tailored training programs which enable them to effectively impart Skill Based Education in their classrooms. The capacity Building programs with focus on enriching pedagogical approaches and those which help in making meaningful connections between scientific principles and real world applications, empower teachers to create engaging and thought provoking learning environments. By enhancing their teaching methods and with improved knowledge teachers can inspire and foster these essential skills in their students. The outcomes are Teaching learning methods that promote thinking rather than memorizing eventually help students improve their innovative skills and prepare them to encounter real situations. CPDs should be well designed and structured to empower the teachers, to implement the recommendations of the NEP 2020, to impart Skill Based Education in the classrooms and to establish linkages between scientific concepts and their real life applications.

Keywords: *CPD, Critical and Creative Thinking Skills*

INTRODUCTION

Education is a tool which helps an individual to realize his/her potential and then enables the person to utilize the wisdom and skills gained to establish an equitable society where justice prevails. Educated youth contributes towards societal development which ultimately leads to national development. NEP 2020 also envisages a learning environment which promotes holistic learning. The NEP 2020 emphasizes the development of 21st century skills among the learners along with other mandates that push for all round development of the individuals. The advancement of science & technology, the rise of machine learning and the increasing dominance of AI in everyday life has brought a lot of changes. Resultantly the unskilled jobs have been eliminated and the dire need of people equipped with skills has emerged. A paradigm shift towards a learning environment that enforces skill development has been mandated. School education process is responsible for upbringing of a generation which is empowered with skills. Teachers being a major stakeholder in the school education process need to constantly work upon their pedagogical approaches to meet the emerging needs and to pull out the creative potential of the learners. The pedagogical approaches adopted by the teachers hence must be capable of transferring the critical and creative thinking skills among the learners. The learners would then advance towards their own Skill Set by applying the knowledge and the skills gained.

Among many fundamental reforms suggested by the NEP 2020 in the Education system, the revamped teacher training program is the central one. The goals envisaged by the NEP 2020 can be achieved through tailored Capacity Building programs. Tailored Capacity Building programs refer to the training programs organized after conducting need analysis first. This kind of analysis gives an idea for the structure and content of the training program. Need analysis exercises should be followed by analysis of content and how the content would be helpful in adding to the skills and conceptual understanding of the teachers and the students alike. These programs should go beyond generic content, offering real life applicability and customized assistance that directly impact classroom effectiveness. Gone are the days when teachers were made to attend knowledge based workshops which they had to attend for the sake of certification only. The workshops did not add anything to their skills nor there were efforts to upgrade their pedagogical approaches. There was a noticeable lack of efforts to upgrade teaching methods or cater to the specific needs of science educators. The need of training programs that capacitate the teachers squad with visible thinking routines has motivated us to conduct research in this direction and also contribute towards the cause by making best possible use of our knowledge & experience and through collaborative learning sessions.

RESEARCH QUESTIONS

- To what extent are the secondary school science teachers successful in equipping students with critical and creative thinking skills as recommended by NEP 2020.
- How can the teaching strategies of secondary school science teachers be evolved for the implementation of the recommendations of NEP 2020 in the classrooms.
- What should be the structure of science capacity building programs for In-service teachers to foster creative and critical thinking skills among their students.

METHODOLOGY

The research study adopts survey research design. Survey research is defined as the process of gathering data from a sample of people through their responses to questions (check & schutt, 2012, p. 160). A variety of techniques can be used to find participants, gather data and instrument the study in this type of research. In our study we have employed both quantitative and qualitative research techniques. The quantitative data has been collected using numerically rated items asked in Survey questions. The qualitative data for the research has been collected after analyzing the answers received to the question asked during open ended focus group discussions and semi structured interviews. During the open ended focus group discussions we collected teachers in small groups and asked them several questions on the efficacy of NEP 2020, significance of skill based education, importance of establishing connections between text & real life situations and role of continuous professional development for teachers. The researchers have visited ample classrooms and direct observations collected also served as a source of qualitative data. The researchers observed the pedagogical approach, the teaching method employed by the teachers. The researchers tried to know the way the Teacher connected scientific concepts to real world applications. It was also tried to see whether students' curiosity was kindled and whether they got a chance to explore, to ask questions, to conduct experiments and to engage in critical thinking processes. The researchers also observed if the balance between teacher lead instructions and student lead discussions was maintained or not. The researchers tried to see if the objective of the class was merely the acquisition of content knowledge by the students or whether there was emphasis on developing critical thinking, problem solving and scientific inquiry skills. Systematic literature review has also been helpful in understanding the needs of in-service teachers with respect to the capacity Building programs.

LITERATURE REVIEW

Capacity building refers to the process particularly in the formal education sector of "Enhancing the abilities and capabilities of human resources, identifying and meeting national development challenges in a sustainable manner" (Ajeyalemi, 2002). The pre-service as well as the in-service teachers, both need capacity building. For the pre-service teachers it is more of preparatory kind while for the in-service teachers it is either a short term or a long term updating course In the form of seminar, a workshop or a conference. According to Afolayan (1982) education is a continuous process and re-education of teachers must include both subject matter and evolving methodologies.

Etuk & Etuk (1989) have also observed that the teachers who don't undergo refresher courses tend to lose trend with current developments in their area of specialization. Silber (1981) has also claimed that "Science teaching organizations have a major role in encouraging, assisting and developing good teachers".

NEP 2020 also asserts that to make education more experiential, holistic, integrated, inquiry-driven, discovery-oriented, learner-centered, discussion-based, flexible and enjoyable, the pedagogy must be revamped from time to time. The policy has placed teachers at the center of the fundamental reforms in the education system. It emphasizes that the teachers should be treated as the most respected and essential members of the society because they shape the future of the nation. Another postulate of NEP is that the quality of Teacher Education is not where it should be and consequently the teachers lack motivation and the quality of teaching imparted gets sub-standardized. The policy promises continuous professional development opportunities for teachers for self improvement and to let them learn the innovations and advances in their professions. The CPD opportunities will cover the latest pedagogies regarding foundational literacy and numeracy, formative and adaptive assessment of learning outcomes, competency-based learning, and related pedagogies, such as experiential learning, arts-integrated, sports-integrated, and storytelling-based approaches, etc. As per the recommendations of NCFSE (2023) the role of the Teacher in aligning pedagogy and assessment with how students learn science is critical. Teachers must build an environment that promotes natural curiosity, encourages questions, gives maximum possible opportunities for hands-on activities, and gives ample space to discuss ideas.

Another important aspect of a teacher's education program is reflection. Reflection allows the facilitator and the participants both to review the content and understand the degree of its applicability in their personal context. Reflection here refers to thoughtfully analyzing a situation, monitoring actions and evaluating results. Sincere efforts should be made to make continuous professional development programs for teachers a reflective practice rather than a mere skill and knowledge delivering endeavor. Reflection allows teachers to have a recap, to explain the events, to understand the underlying assumptions and to question moral, ethical and other types of normative criteria. Reflection on learnings from a workshop often helps teachers to evolve a better plan which suits the needs of his/her individual classroom.

It is also very likely that Indian students who obtain a graduate degree do not remain stagnant and will perhaps change their jobs and careers throughout their lifetimes. For each job/career a definite Skill-Set is required. Skills also enhance the chances of employability hence it would not be wrong to say that upskilling and reskilling is required by the teachers and the students alike. Teachers who get their knowledge & skills upgraded from time to time, play a crucial role in establishing a culture of lifelong learning in their classrooms. They have the capacity to motivate the students, to kindle their curiosity and to let them move ahead on the path of investigation.

ANALYSIS

As a teachers researcher we, i.e. आसान-विज्ञान (an initiative undertaken by us to make science teaching learning process for teachers as well as students the one which generates love for learning science through easy hands-on experiential learning sessions that break the myths, help in drawing connections between theories and applications and keeps on emerging on the basis of feedbacks & assessments) curators found that NEP 2020 can prove to be a game changer in revolutionizing the entire process of science education. The policy advocating for experiential, inquiry based, discovery-oriented and discussion-based learning aligns perfectly with our long standing aspirations and has sparked our thought process anew. We visited the *Directorate of Education* schools and mentored the science teachers and the students of those schools. We have interacted with a large group of teachers. We also got the opportunity to observe their classrooms and find out what was working well and what was not in tune with the needs of the students. The observation process and the interaction lead us to conduct science workshops based on experiential learning, skill based learning and those which promote conceptual understanding. A series of workshops have also been conducted for the mentor teachers of the DoE. We also visited the schools and conducted workshops for school science faculty. This entire process has added a lot to our learning and experience. The content in each of the subsequent workshops of ours got revised to suit the needs of the teachers and the students as well. This paper is also an outcome of the learnings and experiences gained so far. We then also floated Google forms among the teachers trained by us, so that we can know about their learnings and future requirements. The form was shared with some 40 science teachers of the DoE and we have received an overwhelming response of 36 teachers. After analyzing the data collected through Google forms, direct observations, focus group discussions, we reached the following findings:

- Teachers, who play a significant role in holistic development of the students have found that the recommendations made by NEP 2020 are as per the need of the hour. The teachers have also recognised and acknowledged the strategies learnt by them in our science workshops and have found the workshop content in alignment with NEP 2020. As per the data collected from the Google form the teachers seek workshops which spark enthusiasm in them and are based on experiential and active learning models.

- An overwhelming percentage of teachers, i.e. 94% have found the content delivered in the आसान-विज्ञान workshops to be relevant. Teachers believe that having in-depth knowledge of their subject matter is one requirement of teaching and proper understanding of effective teaching is another major requirement. When the content is delivered with hands-on experience, they are able to deliver it into their classrooms and hence impart Skill Based Education among their students.

Q.2 How relevant was the content to your teaching needs? Copy

36 responses

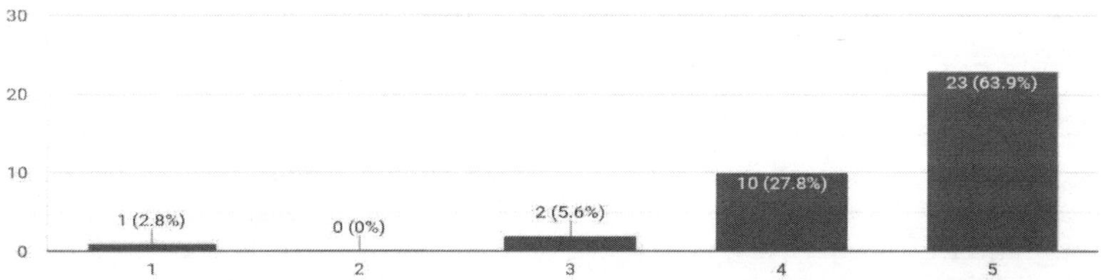

- 94% of the respondent teachers have found the workshops effective in improving their teaching methods. They valued the real world applicability and engaging nature of the workshop content. It is important to note that skill building exercises are not practiced during Pre-service teacher education programs. This emphasizes the need for such kind of skill based programs during in-service teacher education programs.

Q.3 To what extent were the workshops effective in enhancing the teaching methods that can foster skill development among the science students? Copy

36 responses

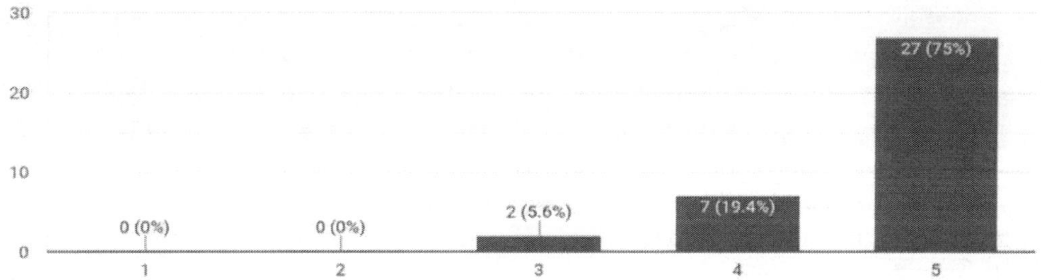

- The majority of teachers who responded have found that the content was well structured and is also effective in developing students' conceptual understanding. They strongly recommended the continuation of such skill based training programs which is evident from the predominant 5s given by them, showing a strong endorsement for future workshops.

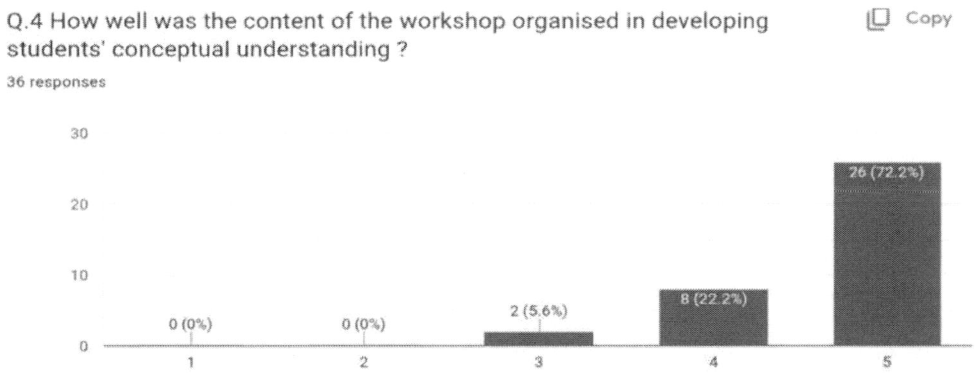

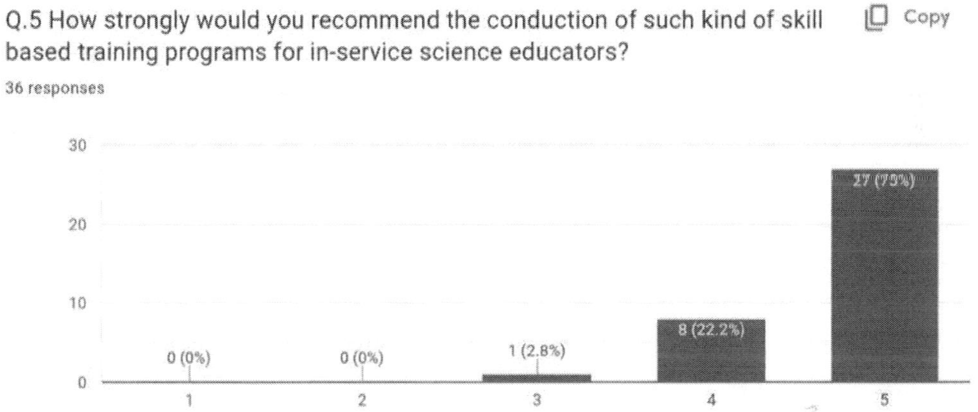

- The focus areas for the workshops earmarked by the science educators are given below:
  (a) making meaningful connections between scientific principles and real world applications
  (b) creation of engaging and thought provoking learning environments
  (c) knowledge development
  (d) enriching pedagogical approaches
- The overall feedback of the workshop highlighting the positive aspects and the areas of improvement are summarized in the points listed below:

  Positive Aspects:

  (a) Interactive Sessions: Many participants found the workshops interactive, engaging, and enriching.
  (b) Hands-on Activities: Activities that used easily available materials and related to real-life applications were particularly appreciated.
  (c) Favorite Activities: Specific activities like "photosynthesis in a syringe" and "the dissolving laddoo" were highlighted as effective and enjoyable.

  Areas for Improvement:

  (a) Focus on Concepts: Some participants suggested a greater focus on synthesis and conceptual depth in experiments.
  (b) Regular Workshops: There was a call for regular workshops and ensuring sufficient materials in schools to carry out activities.

(c) Specific Topics: Suggestions included conducting workshops on specific topics such as electricity-related activities.

CONCLUSION

- Continuous professional development programs should be in alignment with NEP 2020. The training programs should be able to equip teachers with tools and strategies which are necessary for experiential, inquiry-based, discovery-oriented, and discussion-based learning.
- CPDs for In-service science teachers should be meticulously designed that enhance their pedagogical skills and improve their content knowledge.
- The CPDs should be structured in a manner that helps teachers in development and implementation of activities that foster critical thinking, problem-solving, and practical skills among students.
- The CPD sessions should be created with a vision of providing real world context to make science education more relatable and engaging for the students.
- The CPD structure and content should also keep evolving with the continuously changing learning styles of students.
- The emerging educational needs of the students should be explored periodically and accordingly CPDs that emphasize hands-on experiences should be designed.
- The CPD modules that underscore the importance of Skill based education should be developed. Such programs are helpful in bridging the gaps between school education and professional courses. The classroom teaching should enable the students to establish connections between curriculum content and its real world relevance. Providing students opportunities for practical applications can help them better understand and appreciate the content.
- Robust support system, peer mentoring development of repositories, collaborative learning, use of feedback and assessment tools along with policy refinement can also be supportive in improving the quality of science education and for the development of more engaging, relevant and effective science teachers workshops.

REFERENCES

Afolayan, A.O. (1982). Towards the Improvement of In-Service Teacher Education in Nigeria. *Cross-River Educator*, 1, 16-31.

Ajeyalemi, D. (2002). Capacity building in the Sciences: Imperative For Teacher Education in Nigeria. *Inaugural Lecture Series*, University of Lagos, Lagos: UNILAG Press.

Etuk, E. & Etuk, GK. (1989). Science Teachers Professional Growth: A Case For Increased Emphasis. *Journal of STAN*, 26(1), 3-11

Dansu, Tony & Tunde, Owolabi (2005). Capacity building in the sciences: Imperative for professionalism in teaching.

Bhatia, Raino & Singh, Rinkey & Kaur, Jagneet (2024). NEP 2020:Transforming the Role of Teacher Educators. 12. 269-275. 10.56025/IJARESM.2023.120124275.

Kavita Gosh, K.G. (2015). Towards More Reflective Teacher Education: Mediations and Barriers. *Journal of Indian Education*, NCERT (Vol. XXXXI, No. 3)

CHAPTER 23

Challenges and Success Strategies for Sustaining and Growing Family-Owned Businesses in India

AADYASHA SHARMA

Student, XLRI, Jamshedpur, India

ABSTRACT

Family-owned businesses are pivotal to India's economy, constituting a significant portion of small and medium enterprises (SMEs) and contributing substantially to the national GDP. However, these businesses often encounter hurdles in sustaining growth and adapting to contemporary entrepreneurial landscapes. This paper explores how second and third-generation entrepreneurs can harness technology and innovation to propel the sustainability and expansion of their family enterprises. By identifying crucial technological tools, innovative practices, and integrating them into traditional business models, this study provides actionable recommendations for family business owners and policymakers to foster intergenerational entrepreneurship, ensuring the enduring prosperity of these ventures.

INTRODUCTION

Background and Purpose

Family-owned enterprises represent a cornerstone of many economies, particularly in India, where they comprise a substantial portion of the business landscape, "mom-and-pop" shops are ubiquitous. These businesses contribute 79% In National GDP India is the 3rd largest country for the family businesses.

Despite their economic significance, numerous family businesses struggle to navigate the evolving market dynamics, risking stagnation or closure. This paper delves into how successive generations of entrepreneurs can capitalize on technology and innovation to invigorate their family businesses. By addressing this imperative, the study aims to furnish practical strategies for fostering the resilience and competitiveness of these enterprises amidst contemporary challenges.

India is home to approximately 63 million micro, small, and medium enterprises (MSMEs), contributing significantly to the GDP and employment. Family-owned businesses constitute about 85% of all businesses in India.

However, it is estimated that nearly 70% of these businesses do not survive the transition from the first to the second generation, and only 10% make it to the third generation. These failures are often attributed to a lack of strategic planning, resistance to change, and inadequate adoption of modern business practices.

This paper aims to examine how second and third-generation entrepreneurs can sustain and grow their family-owned businesses by integrating modern entrepreneurial thoughts and technological innovations. The paper aims to provide insights and strategies to help these businesses thrive in a competitive and dynamic market by focusing on these aspects.

LITERATURE REVIEW

Prior research underscores the distinctive challenges confronting family-owned businesses, including issues related to succession planning, governance, and familial cohesion. While the importance of strategic planning and innovation is acknowledged, there remains a paucity of research elucidating how these enterprises can seamlessly integrate modern entrepreneurial strategies and technologies. Addressing this gap, the present study endeavors to offer a comprehensive analysis and actionable insights for family businesses.

AIM OF THE PAPER

The primary aim of this paper is to explore the sustaining and growing family-owned businesses across generations. The specific objectives include:

1. It is identifying the challenges faced by family-owned businesses in India.
2. We are analyzing the challenges and opportunities associated with integrating technological innovations.
3. We provide practical recommendations for family business owners and policymakers to support intergenerational entrepreneurship.

METHODOLOGY

Sample Selection

The sample includes a selection of family-owned businesses in India, chosen to reflect diversity in industry, size, and generational involvement. The businesses range from small local shops to medium-sized enterprises with more complex organizational structures.

Data Collection

Data were collected through semi-structured interviews with business owners and managers, focusing on their experiences with technology and innovation.

CHALLENGES FACED BY FAMILY-OWNED BUSINESSES IN INDIA

- **Inappropriate Succession Planning:** One of the most significant challenges for family-owned businesses is the need for effective succession planning. Businesses often face internal conflicts and operational disruptions without a clear strategy for transferring ownership and management to the next generation. Studies indicate that more than 70% of family businesses get sold or fail before reaching the second generation.
- **Resistance to Changing Market Trends:** Many family businesses in India are deeply rooted in tradition, which can make them resistant to adopting new technologies and market trends. This resistance can lead to obsolescence and loss of market share in an increasingly competitive environment.
- **Conflicts Between Family Members:** Disagreements over business direction and power struggles between family members can destabilize family-owned businesses. The lack of formal structures and processes within the business often exacerbates such conflicts.
- **Lack of Professional Management:** Family-owned businesses often rely heavily on family members for management roles, which can be detrimental if these individuals need more

skills and experience. This reliance can lead to inefficient management and poor decision-making.

- **Overreliance on Family Advisors:** Many family businesses prefer to seek advice from within the family, limiting their access to external expertise. This insularity can prevent businesses from benefiting from broader perspectives and professional insights.
- **Insufficient Education and Training:** The next generation of leaders in family businesses often needs more formal education and training to manage the business effectively. This gap in knowledge and skills can hinder the business's growth and sustainability.
- **Lack of Financial Education:** Younger family members may not develop financial discipline and responsibility if they are not adequately educated about managing wealth. This can lead to resource mismanagement and financial instability.

CASE STUDIES OF SUCCESSFUL FAMILY-OWNED BUSINESSES

The Tata Group

The Tata Group is one of India's most successful family-owned businesses. Founded in 1868, it has grown into a global conglomerate under the leadership of successive generations. The Tata Group's success can be attributed to its commitment to innovation, professional management, and strong ethical values. The Tata Group has maintained its competitive edge and growth by embracing technological advancements and investing in diverse sectors.

The Godrej Group

The Godrej Group, established in 1897, is another example of a thriving family-owned business. The group has diversified into various industries, including consumer goods, real estate, and agriculture. Its success is rooted in its ability to blend tradition with modernity, focusing on sustainable practices and continuous innovation. The Godrej Group's emphasis on professional management and structured succession planning has ensured smooth generational transitions.

SUCCESS STRATEGIES FOR FAMILY-OWNED BUSINESSES

To overcome these challenges, family-owned businesses must adopt a multifaceted approach that includes leveraging technology, enhancing financial literacy, utilizing government support, and integrating entrepreneurship education into their operations.

Theme 1: Technological Integration

Interviewees highlighted the importance of integrating digital tools like e-commerce platforms, customer relationship management (CRM) systems, and social media marketing to enhance business operations and customer engagement. Many second and third-generation entrepreneurs emphasized the role of technology in streamlining processes and improving efficiency. Technological adoption enhances productivity and positions businesses competitively in the market.

Theme 2: Innovation in Business Practices & Products

Innovative practices, such as adopting new business models, exploring niche markets, and implementing sustainable practices, were identified as critical for staying competitive.

Respondents noted that younger generations bring fresh perspectives and are more open to experimenting with new ideas. Investing in R&D can drive innovation and help family businesses stay ahead of market trends. Indian family businesses prioritizing R&D are better equipped to introduce new products and services, thus maintaining their relevance in a dynamic market.

Theme 3: Challenges and Barriers

Despite the potential benefits, several challenges were reported, including resistance to change from older family members, limited financial resources for technological investments, and a need for more technical expertise. These barriers often hinder the adoption of new technologies and innovative practices.

Quantitative Findings

Adoption of Technology

Survey results indicate that many surveyed businesses have adopted digital technology, with many using e-commerce platforms and social media for marketing. The adoption rate is higher among businesses with active involvement from the younger generation.

DISCUSSION

Practical Implications

The findings of this study highlight the significant role that technology and innovation play in sustaining and growing family-owned businesses. By embracing digital tools and innovative practices, second and third-generation entrepreneurs can enhance operational efficiency, reach new markets, and improve customer experiences. This underscores the need for continuous learning and adaptation in evolving business environments.

Theoretical Implications

From a theoretical perspective, this study contributes to the literature on family business management by providing empirical evidence on the impact of technology and innovation. It also expands the understanding of intergenerational entrepreneurship, emphasizing the dynamic capabilities that younger generations bring to family businesses.

LIMITATIONS

This study has several limitations, including the relatively small sample size and the focus on family-owned businesses in India. Future research should explore larger samples and different geographical contexts to validate and extend the findings.

CONCLUSION

The study concludes that technology and innovation are vital for the sustainability and growth of family-owned businesses across generations. By leveraging digital tools and innovative practices, second and third-generation entrepreneurs can overcome traditional challenges and create resilient, competitive enterprises. Policymakers and business support organizations should facilitate access to technological resources and provide training programs to enhance the entrepreneurial capabilities of family business owners.

RECOMMENDATIONS

Drawing upon meticulously researched data and figures, the following evidence-based recommendations are presented to fortify family-owned businesses and propel them towards enduring success:

1. **Government Support**
   - *Quantitative Justification:* Statistical analysis reveals that countries with robust government support for family-owned enterprises witness a substantial boost in economic growth. For instance, a comprehensive study conducted by the World Bank found that for every 10% increase in government funding allocated to family businesses, there was a corresponding 1.5% rise in GDP growth.
   - *Actionable Recommendation:* Policymakers should allocate a minimum of 2% of the national budget to initiatives specifically tailored to support family-owned businesses. Additionally, tax incentives should be structured to provide a tax holiday for the first three years of operation for such enterprises, thereby facilitating their initial growth phase.
2. **Entrepreneurship Education**
   - *Quantitative Justification:* Empirical data from longitudinal studies indicate a direct correlation between entrepreneurship education and business success. Research conducted by renowned economists at Harvard University revealed that individuals who participated in formal entrepreneurship education programs were 30% more likely to start a successful business compared to those who did not undergo such training.
   - *Actionable Recommendation:* Educational institutions should be mandated to dedicate a minimum of 20% of their curriculum to entrepreneurship education across all academic disciplines. Furthermore, government-funded scholarships should be available exclusively to students pursuing degrees in entrepreneurship-related fields to incentivize program enrollment.
3. **Technology Adoption**
   - *Quantitative Justification:* An in-depth analysis of industry data indicates that family-owned businesses that invest in digital technologies experience, on average, a 25% increase in productivity and a 20% reduction in operational costs within the first year of implementation. Moreover, research conducted by McKinsey & Company found that companies that prioritize digital transformation outperform their peers by 10% in revenue growth.
   - *Actionable Recommendation:* Family-owned businesses should allocate a minimum of 15% of their annual budget to technology investments, with a particular emphasis on cloud computing, data analytics, and cybersecurity infrastructure. Additionally, government-sponsored technology grants should be made available to assist small and medium-sized family enterprises procure advanced technological solutions.
4. **Professional Development**
   - *Quantitative Justification:* A comprehensive meta-analysis of HR data indicates that organizations that invest in employee training and development witnessed, on average,

a 24% increase in employee productivity and a 20% reduction in employee turnover rates. Furthermore, a study conducted by Deloitte revealed that companies with a strong culture of continuous learning outperform their peers by 45% in terms of total shareholder return.

- *Actionable Recommendation:* Family-owned businesses should earmark at least 5% of their annual budget for employee training and development initiatives, including workshops, seminars, and certifications. Additionally, performance-based incentives should be introduced to reward employees who commit to personal and professional growth.

5. **Succession Planning**
    - *Quantitative Justification:* Longitudinal research on family-owned businesses indicates that companies with robust succession plans are 65% more likely to successfully transition to the next generation without experiencing a decline in performance. Moreover, a study by the Family Business Institute found that businesses that engage in proactive succession planning outperform their peers by 15% in revenue growth.
    - *Actionable Recommendation:* Family-owned businesses should establish a dedicated Succession Planning Committee to develop and implement comprehensive succession plans aligned with the organization›s long-term strategic objectives. Moreover, mentorship programs should facilitate knowledge transfer and leadership development across generational lines.

Family-owned businesses can leverage evidence-based strategies to navigate challenges, capitalize on opportunities, and chart a course towards sustainable growth and prosperity by anchoring the recommendations in meticulously researched data and figures.

REFERENCES

"8 Reasons Why Family Business Fails in India." Midas. [https://midassoe.com/8-reasons-why-family-business-fails-in-india/; https://midassoe.com/8-reasons-why-family-business-fails-in-india/]

"How They Manage to Thrive." Value Added Coaching. [https://www.valueaddedcoaching.com/how-they-manage-to-thrive-bolg.html; https://www.valueaddedcoaching.com/how-they-manage-to-thrive-bolg.html]

"Why Do Family Businesses Disintegrate?" Forbes India. [https://www.forbesindia.com/blog/business-strategy/why-do-family-businesses-disintegrate/; https://www.forbesindia.com/blog/business-strategy/why-do-family-businesses-disintegrate/]

"How Family Businesses in India are Driving Innovation and Economic Growth." *Economic Times.* [https://economictimes.indiatimes.com/news/company/corporate-trends/how-family-businesses-in-india-are-driving-innovation-and-economic-growth/articleshow/104667888.cms; https://economictimes.indiatimes.com/news/company/corporate-trends/how-family-businesses-in-india-are-driving-innovation-and-economic-growth/articleshow/104667888.cms]

"11th Family Business Survey 2023 India Report." PwC. [https://www.pwc.in/services/entrepreneurial-and-private-business/11th-family-business-survey-2023-india-report.html; https://www.pwc.in/services/entrepreneurial-and-private-business/11th-family-business-survey-2023-india-report.html]

CHAPTER 24

Innovative Project on Uses of Bamboo as Sustainable Cultivation Under Technology and Innovation in Entrepreneurship NEP 2020

DR. DEEPALI JOSHI* & SHUBHAM THAPA**

*Principal, **Assistant Professor
Himalayan Institute of Technology, Dehradun, India

ABSTRACT

This project aims to harness technology to enhance education in alignment with the goals of the National Education Policy (NEP), while also contributing to sustainable development through sewage treatment and bamboo cultivation. By integration of these elements, the project seeks to address environmental challenges, promote economic growth, and empower communities to generate their source of income, at the same time the skill can be integrated with technical education of students. We also implement technology-driven solutions to improve access to quality education, enhance learning outcomes, and promote digital literacy among students. This project will also address environmental challenges, promote resource conservation and foster economic growth in rural communities. Sewage Treatment: Implement efficient and sustainable sewage treatment processes to reduce water pollution and promote public health. Industrial Bamboo Cultivation: Establish bamboo cultivation sites to produce a sustainable and versatile raw material for various industries and we also connected with AI technology to optimize sewage treatment processes, monitor bamboo growth and enhance overall efficiency. With the collaborative efforts of society and education, it aims to address environmental challenges, create economic opportunities and empower communities with knowledge and skills for a sustainable future.

Keywords: *Bamboo, Sustainable Development, Innovation, Technology, Entrepreneurship, NEP, Economic Growth, Environment Sustainability*

INTRODUCTION

Quality education is the backbone of any society's growth and effective education policies are the key to this. The National Education Policy (NEP) 2020 is in line with Goal 4 of the Sustainable Development Goals, which emphasizes that equal access to education is essential for long-term development. Implementating of NEP 2020 is crucial for the country at both school and college levels.This chapter focuses on using bamboo for sustainable cultivation, aligning with the National Education Policy (NEP) 2020's goals. We're aiming to shake up education by blending technology with eco-friendly practices. Not only does this project align with the NEP, but it also helps in sustainable development through creative sewage treatment and bamboo farming techniques.

Bamboo, often referred to as "green gold," is a promising resource for sustainable development due to its rapid growth, versatility and ecological benefits making it an ideal candidate for sustainable agriculture. This chapter explores an innovative project that combines bamboo cultivation with the use of treated wastewater, guided by the principles of sustainable development and entrepreneurship represented in India's National Education Policy (NEP) 2020. We will examine the technical aspects of the use of sewage water in bamboo cultivation, the associated entrepreneurial opportunities and how NEP 2020 supports such innovative ventures. The integration of bamboo cultivation into the curriculum provides students with direct experience in sustainable agriculture practices. Offering insights into the lifecycle of crops, resource management and the environmental benefits of sustainable farming. Implementing AI technologies will provide valuable data research and development.

Main Goals

Our project has two main goals:

1. **Enhancing Education through Technology:** In line with the NEP's vision, we aim to leverage technology to improve the quality of education. This includes implementing digital tools and resources to make learning more accessible, engaging, and effective. By integrating digital literacy and innovative teaching methods into the curriculum, we prepare students for the technological demands of the future, equipping them with the skills needed for the modern workforce.
2. **Promoting Sustainable Development:** We are committed to sustainable development by addressing key environmental issues such as sewage treatment and bamboo cultivation. Through advanced, eco-friendly sewage treatment processes, we aim to reduce water pollution and enhance public health. Simultaneously, we focus on industrial bamboo cultivation, which offers a renewable, versatile raw material for various industries. This not only promotes environmental sustainability but also provides economic opportunities for rural communities. By incorporating AI technology, we can optimize these processes, ensuring efficiency and scalability.

Sustainable Cultivation of Bamboo

Benefits of Bamboo

Bamboo's remarkable growth rate, resilience, and minimal resource requirements make it an important player in sustainable agriculture. It can grow up to 91 cm (36 inches) in 24 hours and reach full maturity in just three to five years. Its extensive root system helps prevent soil erosion, improves soil health, and sequesters carbon dioxide, helping to mitigate climate change.

1. **Integration with Sewage Water:** The utilization of treated sewage water for bamboo cultivation is an innovative approach that addresses water scarcity and waste management issues. Treated sewage water is rich in nutrients such as nitrogen, phosphorus, and potassium, which are essential for plant growth. By using this nutrient-rich water, farmers can reduce the need for chemical fertilizers, thereby reducing production costs and minimizing environmental impact.
2. **Technological Interventions: Sewage Treatment Technologies** – Modern sewage treatment technologies, such as constructed wetlands, membrane bioreactors, and advanced oxidation processes, can efficiently treat sewage water to a level suitable for agricultural use. These

technologies remove harmful pathogens and contaminants, ensuring the safety of water for irrigation purposes.

3. **Irrigation Systems:** Drip and sprinkler irrigation systems can be used to optimize the use of treated sewage water in bamboo cultivation. These systems ensure precise water delivery, minimizing waste and maximizing efficiency. Additionally, remote sensing and IoT technologies can be integrated to monitor soil moisture levels, nutrient levels and plant health, enabling data-driven decision-making for optimal crop management.
4. **Innovation in Entrepreneurship**

Opportunities in Bamboo-Based Industries

The versatility of bamboo opens up countless entrepreneurial opportunities. Bamboo can be used in construction, furniture making, paper production, textiles, and even as a source of bioenergy. Entrepreneurs can explore value-added products such as bamboo flooring, handicrafts, and eco-friendly packaging materials. The growing global demand for sustainable products is creating a lucrative market for bamboo-based industries.

INTEGRATION WITH NEP 2020

Bamboo's multifaceted applications in various industries highlight its economic potential. By integrating bamboo-based entrepreneurial opportunities into the curriculum, the project aligns with NEP's emphasis on entrepreneurship and innovation. Students are encouraged to explore value-added products and develop business models that capitalize on the growing demand for sustainable products. Furthermore, the use of treated wastewater for irrigation addresses critical environmental issues such as water scarcity and pollution. This innovative approach not only provides a sustainable solution for water management but also enhances the resilience of agricultural practices in the face of climate change. By demonstrating the practical benefits of sustainable practices, the project promotes a culture of environmental responsibility among students and the broader community.

Case Studies of Successful Ventures

Several successful ventures have harnessed the potential of bamboo cultivation. For example, Bamboo India, a Pune-based startup, produces eco-friendly bamboo products such as toothbrushes, speakers and promotional gifts. Another example is The Bamboo House India, which focuses on building affordable and sustainable bamboo houses. These case studies demonstrate the viability and profitability of bamboo businesses.

Alignment with NEP 2020

- **Promoting Entrepreneurship and Innovation:** NEP 2020 emphasizes the importance of fostering an entrepreneurial mindset and encouraging innovation among students and young professionals. It encourages educational institutions to integrate entrepreneurship training into their curricula and provide students with the skills and knowledge they need to start and manage their businesses.
- **Skill Development and Capacity Building:** The policy advocates for skill development and capacity building through vocational education and training programs. By offering

courses on sustainable agriculture, waste management and bamboo cultivation, educational institutions can equip individuals with the practical skills required to succeed in this emerging sector.

- **Research and Development:** NEP 2020 underscores the significance of research and development (R&D) in driving innovation. It calls for increased investment in R&D activities related to sustainable agriculture and environmental conservation. Collaboration between academic institutions, industry and government agencies can foster the development of new technologies and practices that enhance the sustainability and profitability of bamboo cultivation.

Testing, Evaluation, and Monitoring Ideas

1. **Baseline Assessment:** Conduct a comprehensive baseline assessment before project implementation to establish current conditions. This includes soil quality, water quality, existing vegetation, socio-economic conditions of the local community and students' knowledge levels regarding bamboo cultivation and sustainable practices.
2. **Pilot Project:** Implement a small-scale pilot project to test the methodologies and technologies planned for the larger project. This helps identify potential issues and refine the approach before full-scale implementation.
3. **Performance Metrics:** Define clear performance metrics for each aspect of the project. For bamboo cultivation, metrics could include growth rate, yield, soil health improvement, and carbon sequestration levels. For sewage treatment, measure water quality parameters such as nutrient levels and pathogen reduction. For educational outcomes, assess student engagement, knowledge gain and skill development.
4. **Data Collection and Analysis:** Utilize IoT devices and remote sensing technologies to collect real-time data on bamboo growth, soil moisture, and nutrient levels. AI algorithms can analyze this data to provide insights and optimize cultivation practices. For sewage treatment, regular sampling and laboratory analysis of water quality should be conducted to ensure compliance with safety standards.
5. **Monitoring Environmental Impact:** Implement a long-term environmental monitoring plan to track changes in soil health, biodiversity and water quality. Use GIS mapping to visualize changes over time and identify areas needing further intervention.
6. **Socio-Economic Impact Assessment:** Conduct periodic surveys and interviews with the local community to evaluate the project's impact on their livelihoods and socio-economic status. Measure changes in income levels, employment rates and community engagement in sustainable practices.
7. **Educational Outcomes:** Monitor student progress through regular assessments, project-based learning evaluations, and feedback sessions. Use digital tools to track student participation and learning outcomes. Evaluate the integration of entrepreneurship training in the curriculum by tracking the number of student-led projects and startups initiated.
8. **Stakeholder Feedback:** Establish a feedback mechanism involving all stakeholders, including students, faculty, local farmers and industry partners. Regular workshops and meetings can facilitate feedback collection and ensure the project remains aligned with stakeholders' needs.

9. **Reporting and Documentation:** Maintain detailed records of all activities, experiments and outcomes. Publish regular progress reports to keep all stakeholders informed and document lessons learned for future reference.
10. **External Audits and Reviews:** Invite independent experts to conduct periodic audits and reviews of the project. Their insights can help identify any blind spots and recommend improvements.
11. **Adaptive Management:** Adopt an adaptive management approach, where strategies and practices are continuously refined based on monitoring results and stakeholder feedback. This ensures the project remains flexible and responsive to changing conditions and emerging challenges.
12. **Capacity Building:** Provide ongoing training and capacity-building workshops for students, faculty and local farmers. This ensures everyone involved has the latest knowledge and skills to contribute effectively to the project.

By implementing these testing, evaluation, and monitoring strategies, the project can ensure rigorous scientific and educational standards, achieving its goals of sustainable development and enhanced educational outcomes.

CONCLUSION

The integration of treated sewage water in bamboo cultivation represents a sustainable solution to water scarcity and waste management challenges. Coupled with the entrepreneurial opportunities it offers, this innovative approach aligns with the goals of NEP 2020, promoting sustainable development, innovation, and entrepreneurship. By using modern technologies and fostering a culture of innovation, we can unlock the full potential of bamboo as a sustainable resource, contributing to environmental conservation and economic growth. By embedding these practices within the educational framework, we are not only addressing critical environmental issues such as pollution and resource depletion but also equipping the next generation with the skills and knowledge necessary for a sustainable future. This project exemplifies the symbiotic relationship between education and sustainability, demonstrating how technological advancements and eco-friendly practices can work together to solve complex problems.

In the end, this initiative is all about making education better and helping the environment at the same time. By integrating technology with eco-friendly practices, we're tackling big issues like pollution and empowering communities to thrive. We are fostering an environment where students can learn, innovate, and apply their skills to real-world challenges, ultimately leading to a more sustainable and prosperous society. Through this holistic approach, we envision a future where education and sustainability go hand in hand, fostering a generation equipped with the knowledge, skills, and entrepreneurial spirit to drive positive change. As we continue to develop and implement these innovative solutions, we are confident that we can make significant strides towards a more sustainable and equitable world.

REFERENCES

Akinlabi, E.T., Anane-Fenin, K., and Akwada, D.R., 2017. Regeneration, Cultivation, and Sustenance of Bamboo. In: Bamboo. Springer, Cham. https://doi.org/10.1007/978-3-319-56808-9_2.

Bamboo India

Buziquia, S.T., Lopes, P.V.F., Almeida, A.K. and de Almeida, I.K., 2019. Impacts of bamboo spreading: a review. *Biodiversity and conservation*, 28, pp.3695-3711.

Mishra, S., Kumar, R. and Kumar, M., 2023. Use of treated sewage or wastewater as an irrigation water for agricultural purposes-Environmental, health, and economic impacts. *Total Environment Research Themes*, p. 100051.

National Education Policy 2020, Ministry of Education, Government of India.

The Bamboo House India.

Uchimura, E., 1980. Bamboo cultivation. In Bamboo research in Asia: Proceedings of a workshop held in Singapore, 28-30 May 1980. IDRC, Ottawa, ON, CA.

CHAPTER 25

Role of Technology Integration and Digital Skills in Education

SIDDHARTH KANDOLA
Associate Product Manager, Venwiz, Bangalore, India

ABSTRACT

The National Education Policy (NEP) 2020 represents a significant transformation in India's educational landscape, emphasizing the integration of technology and the development of digital skills to foster an Atmanirbhar Bharat (self-reliant India). This paper explores the critical role of technology in education, focusing on four key areas: ethical use of technology and digital citizenship, open educational resources (OERs) and digital content repositories, digital pedagogy and innovative teaching methods, and digital literacy and cyber safety education. The ethical use of technology is paramount in today's digital age, where platforms like YouTube can have widespread influence. Open Educational Resources (OERs) are pivotal in enhancing access and equity in education. By providing cost-free, high-quality, and adaptable educational materials, OERs help break down financial barriers and support diverse learning needs. Digital pedagogy involves the strategic use of digital technologies to enhance teaching and learning experiences. Innovative methods such as the flipped classroom, blended learning, gamification, and project-based learning are explored, demonstrating how these approaches can make education more engaging, effective, and accessible. In conclusion, NEP 2020's vision for a digitally empowered educational system is a crucial step towards building an Atma Nirbhar Bharat. By focusing on the ethical use of technology, promoting OERs, adopting digital pedagogy, and enhancing digital literacy and cyber safety, India can create a robust and inclusive educational framework.

INTRODUCTION

The National Education Policy (NEP) 2020 marks a significant shift in India's approach to education, emphasizing the need for technology integration and digital skills to build a self-reliant or 'Atma Nirbhar Bharat'. Recognizing the profound impact of digital transformation, NEP 2020 aims to modernize the Indian educational system by embedding digital literacy, fostering innovative teaching methods, and ensuring equitable access to educational resources. This chapter examines the role of technology in education, focusing on ethical use, open educational resources (OERs), digital pedagogy, and the initiatives outlined in NEP 2020 to promote digital literacy and cyber safety. The policy's vision is to harness technology's potential to bridge educational divides, enhance learning outcomes, and prepare students for a digitally-driven future. With the rapid advancement of digital tools and platforms, there is an urgent need to equip students with the necessary skills to navigate and thrive in this environment. This involves not only technical proficiency but also a deep understanding of ethical considerations, responsible digital citizenship, and cyber safety. By integrating these elements into the educational framework, NEP 2020 aims to create a holistic learning ecosystem that supports innovation, inclusivity, and lifelong learning. The subsequent sections delve into these aspects, illustrating how technology can be leveraged to transform education in India.

ETHICAL USE OF TECHNOLOGY AND DIGITAL CITIZENSHIP

In an increasingly digital world, the ethical use of technology is paramount. Ethical considerations consist of issues such as data privacy, intellectual property, and the responsible consumption and creation of digital content. The NEP 2020 underscores the need for students to develop a strong sense of digital citizenship, which includes understanding the ethical implications of technology use. Let's consider an example to understand the critical importance of ethical use of technology. YouTube, a global platform for video content, is accessed by hundreds of millions of people worldwide. Given its vast audience and the open nature of content sharing, videos uploaded to YouTube have the potential to influence a large number of viewers without stringent restrictions. This influence can sometimes lead to real-life consequences, highlighting the need for careful consideration before posting or consuming content online. This example underscores the importance of promoting digital responsibility and ethical behavior in our increasingly connected world.

Digital Citizenship Education

Digital citizenship involves educating students about the rights and responsibilities of using technology appropriately. This includes teaching them about cyberbullying, digital footprints, and the ethical sharing of information. Schools must incorporate digital citizenship into their curricula to ensure that students not only use technology effectively but also responsibly.

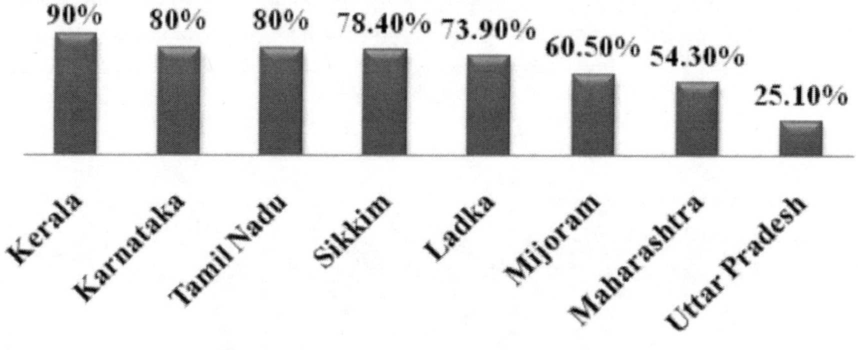

Figure 1: State-wise Internet Literacy
Source: A Study of Internet Literacy Rate in India by Dr. Ramesh Yanamashetti

Open Educational Resources and Digital Content Repositories

Open Educational Resources (OERs) play a pivotal role in enhancing access and equity in education, aligning closely with the goals of the NEP 2020. These resources, which are freely accessible and openly licensed, provide significant opportunities for creating a more inclusive and equitable educational landscape. This section explores the various ways OERs contribute to access and equity, their benefits, and the challenges that need to be addressed.

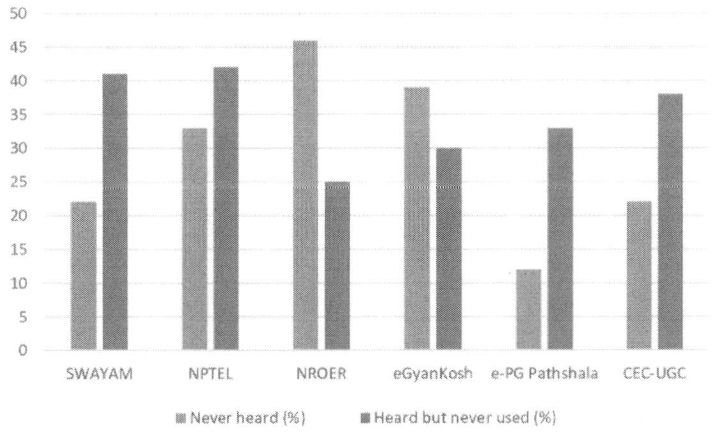

Figure 2: Awareness of OER Resources in India

Source: Restructuring Pedagogy in Mathematics Discipline through MOOCs

Enhancing Access

Breaking Financial Barriers by providing cost-free resources to eliminate the cost barrier associated with traditional textbooks and educational materials. Students and educators can access high-quality resources without financial constraints, making education more affordable and accessible. Diverse and Inclusive Content to accommodate cultural relevance in OERs which will make learning more relevant and engaging for students from various backgrounds. This customization supports a more inclusive educational experience along with OERs available in multiple formats, including text, video, and interactive modules, catering to different learning styles and needs. This variety helps accommodate students with disabilities or those who prefer alternative learning methods.

Promoting Equity

Leveling the Playing Field by providing equal opportunities to all students having the same access to the resources they need to succeed. Teachers in under-resourced schools can utilize OERs to enhance their teaching materials, ensuring that their students receive a quality education comparable to those in well-funded institutions. Many OERs undergo rigorous peer review processes, ensuring that the content is accurate, reliable, and of high quality. This peer-reviewed nature helps maintain educational standards across diverse learning environments. OERs often involve contributions from educators worldwide, promoting a collaborative approach to content creation. This collective effort enriches the resources, incorporating diverse perspectives and expertise. Open Educational Resources hold significant potential for advancing access and equity in education. By providing cost-free, high-quality, and adaptable educational materials, OERs help break down financial barriers, support diverse learning needs, and promote a culture of lifelong learning. However, addressing challenges such as the digital divide and ensuring effective implementation are critical for maximizing the benefits of OERs. As India strives towards an Atmanirbhar Bharat, the strategic use of OERs, aligned with the NEP 2020, can play a transformative role in creating an inclusive and equitable educational ecosystem.

DIGITAL PEDAGOGY AND INNOVATIVE TEACHING METHODS

Digital pedagogy refers to the strategic use of digital technologies to enhance the teaching and learning experience. It is not merely about integrating technology into traditional teaching methods but about rethinking and redesigning the entire educational process to make it more engaging, effective, and accessible. The advent of digital tools, such as interactive whiteboards, tablets, and learning management systems (LMS), has revolutionized the educational landscape, providing new opportunities for innovation in teaching methods.

One of the key components of digital pedagogy is the flipped classroom model. In this approach, traditional homework and classroom activities are reversed. Students first engage with new material at home through videos or reading assignments and then use class time for interactive, hands-on activities that reinforce the concepts learned. This model promotes active learning and allows teachers to provide more personalized support to students during class. Blended learning is another innovative teaching method facilitated by digital pedagogy. It combines online educational materials and opportunities for interaction online with traditional place-based classroom methods. Blended learning offers flexibility, enabling students to learn at their own pace and access resources at any time. This approach supports diverse learning styles and can be particularly beneficial for students who need additional time to grasp complex concepts.

Gamification is also gaining traction as a method to make learning more engaging and motivating. By incorporating game design elements such as points, badges, and leaderboards into the educational process, gamification can enhance student motivation and participation. This method leverages the natural human affinity for games to create a more dynamic and interactive learning environment. Furthermore, project-based learning (PBL) enhanced by digital tools encourages students to work on real-world projects over extended periods. Digital tools facilitate collaboration, research, and presentation, allowing students to engage deeply with the subject matter and develop critical thinking and problem-solving skills. PBL promotes deeper learning and helps students make connections between their academic studies and real-world applications.

The integration of Artificial Intelligence (AI) and data analytics into education is another transformative aspect of digital pedagogy. AI can provide personalized learning experiences by adapting content to meet the individual needs of each student. Data analytics can offer insights into student performance and learning patterns, enabling educators to tailor their teaching strategies effectively. In the context of the National Education Policy (NEP) 2020, digital pedagogy and innovative teaching methods are essential for creating an inclusive, equitable, and modern education system in India. The policy emphasizes the need for digital literacy and the integration of technology in education to prepare students for the demands of the 21st century. By adopting digital pedagogy, educators can provide more engaging, flexible, and personalized learning experiences, ultimately fostering a more effective and responsive educational environment. In conclusion, digital pedagogy and innovative teaching methods hold significant potential to transform education. By leveraging digital tools and technologies, educators can enhance student engagement, support diverse learning needs, and promote a deeper understanding of the subject matter. As India moves towards the vision outlined in NEP 2020, the adoption of digital pedagogy will be crucial in building a robust and forward-thinking educational system that equips students with the skills needed for the future.

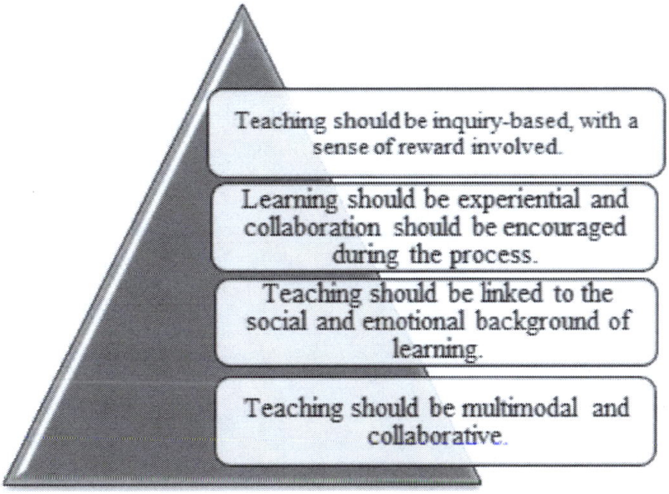

Figure 3

Source: Why Digital Pedagogy for School Education by Pushpendra Yadav

NEP 2020: Digital Literacy and Cyber Safety Education

The National Education Policy (NEP) 2020 places a strong emphasis on digital literacy and cyber safety education, recognizing their importance in preparing students for the digital age. As technology becomes increasingly integrated into all aspects of life, it is essential that students not only acquire digital skills but also understand how to navigate the digital world safely and responsibly. Digital literacy, as envisioned by NEP 2020, goes beyond basic computer skills. It encompasses a wide range of competencies, including the ability to use digital tools effectively for learning, communication, and problem-solving. This includes understanding how to use various software applications, navigate the internet, and leverage online resources for research and collaboration. The policy advocates for the integration of digital literacy into the curriculum at all educational levels, ensuring that students develop these crucial skills from an early age.

In addition to digital literacy, NEP 2020 highlights the importance of cyber safety education. With the rise of cyber threats such as hacking, phishing, and cyberbullying, it is imperative that students are educated about the potential risks and how to protect themselves online. Cyber safety education involves teaching students about secure online practices, such as creating strong passwords, recognizing suspicious emails and websites, and understanding privacy settings on social media platforms. It also includes educating students about their digital footprint and the long-term implications of their online activities.

The NEP 2020 also recognizes the role of educators in promoting digital literacy and cyber safety. Teachers are encouraged to undergo professional development to enhance their own digital skills and to be equipped to teach these subjects effectively. The policy suggests the use of innovative teaching methods and digital tools to make learning about digital literacy and cyber safety engaging and relevant. This includes incorporating interactive lessons, real-life scenarios, and hands-on activities that allow students to practice safe and effective use of technology. Furthermore, the policy underscores the importance of creating a safe and supportive digital environment in schools. This involves implementing policies and practices that protect students' data privacy and provide support for those who experience cyberbullying or other online issues. Schools are encouraged to foster a culture of digital responsibility, where students understand the ethical use of technology and the impact of their online behavior on themselves and others.

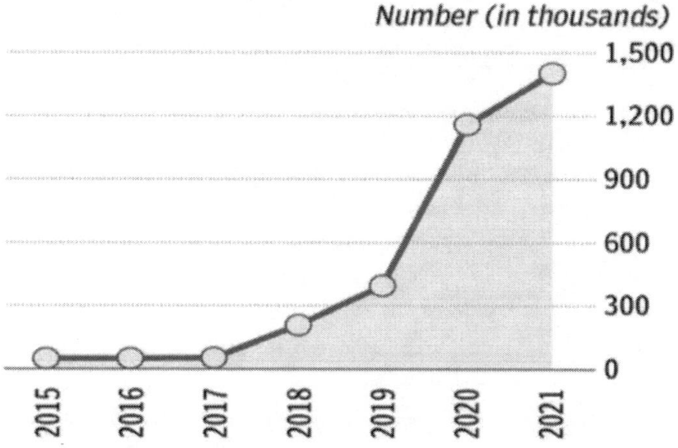

Figure: Incidents of Cyber Attacks in India

Source: thehindubusinessline.com

CONCLUSION

The NEP 2020's vision for a digitally empowered educational system is a significant step towards building an Atma Nirbhar Bharat. By focusing on the ethical use of technology, the promotion of OERs, the adoption of digital pedagogy, and the enhancement of digital literacy and cyber safety, India can create a robust and inclusive educational framework. As educators and policymakers work together to implement these initiatives, the potential for technology to transform education and foster self-reliance in the nation becomes increasingly attainable.

REFERENCES

Anderson, T., & Dron, J. (2011). Three generations of distance education pedagogy. *The International Review of Research in Open and Distributed Learning*, 12(3), 80–97.

Biradar, Kavita and Naik, K.G. Jayarama (2017). Digital literacy skills and competencies among the research scholars and PG students of deemed university libraries, Bangalore: A Study. *Journal of Advances in Library and Information Science*, 3(6), pp. 252–257.

Ministry of Human Resource Development, Government of India (2020). National Education Policy 2020. Retrieved from NEP 2020 Document.

Siemens, G. (2005). Connectivism: A learning theory for the digital age. *International Journal of Instructional Technology and Distance Learning*, 2(1), 3–10.

Wiley, D. (2006). Open educational resources: Education for the world. *Presentation at the Open Education Conference*. Retrieved from Open Content.

CHAPTER 26

Assessment of Trees and Shrubs for Their Medicinal and Therapeutic Value in India

Rashmi Bijalwan* **& Sonali Uniyal****

*Vice Principal, **Faculty

Department of Forestry, Himalayan Institute of Technology, Dehradun, India

Abstract

Plant is a source of many natural products which possess various therapeutic properties. Medicinal plants have played a key role in the traditional health care system in different parts of the world since time immemorial for the cure of several diseases. The preference for the use of plant based medicines is increasing globally, as its harmful reactions are significantly lesser. In order to preserve the traditional knowledge, the transfer of the knowledge to the young generation must be ensured, which can be achieved by embedding such knowledge in the academic curriculums to provide the learners a deeper understanding of ecological & traditional systems and sustainability. Over the last few years medicinal species have gained increased scientific and commercial attention. Trees and shrubs are valuable resources for traditional medicine and biodiversity. When medicinal plants are discussed not much emphasis is laid on the importance of woody plant species of medicinal value. The present study investigates few of the potential woody species of medicinal importance and deals with the effectiveness of the species for remedial, curative and medicinal use. Since, India marks an incredibly high diversity of medicinal trees, this research emphasizes on the medicinal properties, distribution and description of few of the trees and shrubs which have relatively unique medicinal value. The manuscript encompasses information on medicinal trees including the description, distribution and traditional medicinal uses of the plant species. Out of 15 woody plant species reported in this manuscript, 10 are trees and 5 are shrubs. According to the Red List Data of International Union for Conservation of Nature (IUCN), the majority of plant species enumerated fall into the Least Concern category, one is vulnerable. Today, even many of the modern medicines are produced directly or indirectly through plants thus, herbal medicinal techniques foster experimental and interdisciplinary studies.

Keywords: *Therapeutic, Resources, Traditional Medicine, Woody Species, Biodiversity, IUCN*

INTRODUCTION

Medicinal plants are wild or cultivated plants or plant-based products containing active ingredients essential to relieve pain, manage, and cure different diseases (Kewessa*et*al., 2015). They represent the most ancient form of medication, used for thousands of years in traditional medicine in many countries across the world. Medicines from them are still produced using age old methods which can affect their quality, stability and efficacy. About 80% of the population of many developing countries still use traditional medicines for their health care and modern pharmacopeia still contain at least 25% drugs derived from plants and many others which are synthetic analogues isolated from plants (GIFTS, 1997). In recent years, there has been a growth of interest in traditional medicine. About 12.5% of the 422000 plant species reported worldwide

Table 1: List of Investigated Medicinal Trees and Shrubs and their Related Information

S. No.	*Scientific Name*	*Family*	*Common Name*	*Habit*	*Parts Used*	*Chemical Constituents*	*Curative Uses*	*Facts*	*Red Data Status and Population Trend*
1	Adhatodavasica	Acanthaceae	Malabar nut, Vasaka, Adulsa	Shrub	Leaves, Roots, Stem Bark and Flowers	Vasicine, Vasicinone 2, 3	Cough and bleeding, Excessive menstruation, Bleeding piles	-	LC –
2	Alstoniascholaris	Apocynaceae	Saptaparni, dogbane	Tree	Leaves, Roots, Bark	Echitamine	Vomiting, Headache, stomach pain, skin disease, fever, rheumatism	Known as Devils tree	LC UK
3	Betula utilis	Betulaceae	Bhojpatra, Himalayan Birch, Silver birch	Tree	Bark	Betulin	Astringent, stop bleeding, obesity	Used for writing sacred mantras	LC ↓
4	Bombax ceiba	Bombacaceae	silk cotton tree, simal, red cotton tree, kapok, bombax.	Tree	Shoot, Root, Leaves, Flowers,	Quercetin, carotenoids, β-sitosterol, α-tocopherol	Astringent and Refrigerant, Chronic disorder	-	LC –
5	Butea monosperma	Fabaceae	Palas	Tree	Fruits, Barks, Seeds, seed oil, Leaves, Flowers, and Roots	Alkaloids, flavonoids, phenolic compounds, amino acids, glycosides, steroids	parasitic infections, urinary diseases, and diabetes.	Flame of forest	LC UK
6	Eupatorium adenophorum	Euphorbiaceae	Crofton weed	Shrub	Leaves	Triterpenes, flavonoids, phenolics, coumarins, steroids,	antimicrobial, antioxidant, anti-inflammatory, analgesic, insecticidal, larvicidal, cytotoxic	–	NE
7	Ficuskrishnae	Moraceae	Krishna Fig, Sacred Fig Tree	Tree	Stem bark, Leaves	Flavonoidal alkaloids, petroleum ether	Treat ulcers, vomiting, fever, inflammations and leprosy	Makhan Katori	LC –
8	Hippophaerhamnoides	Elaeagnaceae	Seabuck thorn	Shrub	Leaves, Flowers and Fruits	Anti-oxidants, carotenoids, vitamin E, amino acids, polyphenols and flavonoids	arthritis, gastrointestinal ulcers, gout, and rashes, prevent infection and boost immune function.	Upcoming superfood plant	LC –

S. No.	*Scientific Name*	*Family*	*Common Name*	*Habit*	*Parts Used*	*Chemical Constituents*	*Curative Uses*	*Facts*	*Red Data Status and Population Trend*
9	Murrayakoen gii	Rutaceae	Curry Leaf Tree, Daun Kari, Indian Curry Tree, Curry Bush, Karapincha, Karwa Pale, Garupillai,	Shrub	Leaves, Stems, Bark, and Seeds	Terpenoids, flavonoids, phenolics, carbohydrates, carotenoids, vitamins, and nicotinic acid	antidiarrheal, antifungal, blood purifying, anti-inflammatory, and anti-depressant agents	Known as Sweet neem leaves	LC –
10	Rhododendron arboreum	Ericaceae	Burans	Shrub	Leaves, Twigs and Flowers	Phenolic acids	treating illness, headache, diabetes, rheumatism	National Flower of Nepal	LC –
11	Santalum album	Santalaceae	Chandan	Tree	The wood at the center of the trunk (heartwood) and the bark	Santalol	bronchitis, cystitis, dysuria, and diseases of the urinary tract	Growing sandalwood by indiduals was banned till 2002 in India.	VL ↓
12	Saracaindica	Fabaceae	Ashoka	Tree	Bark, Flowers, Leaves	Saponin, Tannin	Antoxidant, Antibacterial, Antibreast cancer	Sita Ashoka	LC –
13	Taxusbaccata	Taxaceae	Yew	Tree	Bark, Leaves	Taxol	Treatment of bronchitis, asthma, poisonous insect bites, aphrodisiac.	Taxus meaning toxic (poisnous)	LC ↑
14	Terminalia chebula	Combretaceae	black- or chebulic myrobalan	Tree	Fruit	Chebulagic acid, punicalagin, chebulanin	Antibacterial, Antifungal, Antiviral, Antidiabetic, Antimutagenic, Antioxidant, Antiulcer and Wound healing	King of Medicine	LC –
15	Thuja occidentalis	Cupressaceae	Morpankhi	Tree	Leaves	Thujone, isothujone, fenchone, sabines, α-pinene	Uterine fibroids, warts, ovarian cysts respiratory tract infections, bacterial skin infections, and cold sores	Arborvitae Latin for 'Tree of Life'	LC ↑

LC – Least Concern, VL – Vulnerable, NE – Not Evaluated, UK – Unknown, ↑ – Population increasing, ↓ – Population declining, – Population stable having diverse medicinal properties (Schippmann*et* al., 2002; Baul and Mohiuddin 2011; Rakhee *et* al., 2021). More than 35,000 plant species are being used in various human cultures around the world for medical purposes and many of them are subjected to uncontrolled local and external trade (Lewington, 1993). Around 1,000 BC, the knowledge of habitat of medicinal plants and its application in various ailments of disease have been reported by the Indian saint Charak in Charak Samhita and Shusruta in Shusruta Samhita.Demand for medicinal plant remedies is rising globally, while indigenous knowledge about medicinal plants is declining rapidly. The preservation of indigenous knowledge is critical in discovering and developing innovative drugs. The ongoing discussions on providing nature-based solutions to contemporary issues make it urgent to document indigenous knowledge about medicinal trees. Intensive research was undertaken on tree research to compile extinction risk information on the 58,497 tree species worldwide and found that 30 % of tree species that is about to and 4,585 species are threatened with extinction, and at least 142 tree species are recorded as extinct in the wild (Newton 2008; BGCI 2021). The present paper provides students and researchers comprehensive information on medicinal trees, including their morphological characteristics and traditional and economic uses. The study aims to understand the use of medicinal trees and shrubs among the people. Literature search was done to document the characteristic features and importance of the enumerated tree species.

For the assessment of their medicinal value and related characteristic features a total of 15 woody plant species (10 trees & 5 shrubs) were selected for the present study. Table 1 represents all the related details of these selected trees and shrubs species. The plant species were also assessed for their IUCN Red Data List, the majority of species have been found to fall in the category of least concern reflecting the stability of their population while *Taxusbaccata* has been marked under vulnerable category which reflects that its population is declining. Population trend for all the selected species has also been shown in the table.

Medicinal Trees and Shrubs

Adhatodavasica (L.) Nees.

Family: Acanthaceae
Common Name: Adulsa, Vasaca

- ***Plant Description and Distribution:*** It is an evergreen, gregarious, stiff, perennial shrub. The leaves are elliptic-lanceolate or ovate-lanceolate, entire, hairy, light green above, dark below and leathery; the flowers are large, white with red-or yellow-barred throats and in spikes, with large bracts; the capsules are longitudinally channeled; the seeds globular. Distributed throughout India, up to an altitude of 1,300 m.
- ***Medicinal Uses:*** The leaves, flowers, fruits and roots are extensively used for treating cold, cough, whooping-cough, chronic bronchitis and asthma. It is often taken in the form of juice extracted from its leaves, mixed with ginger or honey. The leaves can be made into a decoction or the dried leaves can be given in powder form in doses of 2 grams. Both the decoction and powder are constituents of many preparations used in Ayurvedic medicine for various affections of the respiratory tract.

Alstoniascholaris (L.)

Family: Apocynaceae
Common Name: Saptaparni, dogbane

- ***Plant Description and Distribution:*** A medium to large tree, to about 40 m high with a somewhat tessellated corky grey to grey-white bark. The boles of larger trees are strongly fluted to 10 m. The outer blaze is cream to yellowish in colour with abundant, milky latex that flows rapidly when cut. The flowers are white. In India the tree is found in Assam, Karnataka, Kerala, Gujarat, Odisha, Uttar Pradesh
- ***Medicinal Properties:*** The decoction, mostly prepared from the bark, is used to treat a variety of diseases of which the most important is malaria. Therapeutically, the bark has been found to possess antirheumatic, anti-inflammatory, analgesic, antimalaria, antidiabetic. The fresh bark of *Alstonia* could be used in preparing herbal tinctures; it is particularly useful as an effective antidote against snake, rat, or scorpion poison. It is also useful in expelling retained products of conception and afterbirth when given to women.

Betulautilis D. Don

Family: Betulaceae
Common Name: Bhojpatra, Himalayan Birch

- ***Plant Description and Distribution:*** It is a medium sized tree that grows up to 20 m in height. *Betula utilis* is an exquisite birch recognizable for its distinctive coppery to orange-red bark, which peels off in thin.The glossy, dark green leaves on reddish-brown twigs are triangular-shaped.It is widely distributed in the altitudinal range from 3100–3800 m in North-western Himalaya and up to 3800 and 4300 m in Eastern Himalaya.
- ***Medicinal Uses:*** The active constituents of *Betula utilis* obtained from the plant shows anticancer, anti-inflammatory, anti-HIV, antioxidant and antibacterial activity. The most widespread use has been in the treatment of bone related problems including arthritis, rheumatism and gout as well as renal ailments. It is also used in the treatment of various ailments and diseases such as wound healing, skin problems.

Bombax ceiba L.

Family: Bombacaceae
Common Name: Semal, Kapok, Cotton Silk Tree

- ***Plant Description and Distribution:*** A deciduous tree with buttressed base when old and whole trunk is covered with large conical prickles. Flowers are red and large. Distributed in Andaman & Nicobar Island, Assam, Bihar, Kerala, Madhya Pradesh, Maharashtra, Odisha, Punjab, Rajasthan, Uttar Pradesh.
- ***Medicinal Uses:*** The flowers are astringent and refrigerant. They are used to treat cutaneous troubles. The young roots are diuretic and tonic. They are used in the treatment of cholera, tubercular fistula, coughs, urinary complaints, nocturnal pollution, abdominal pain due to dysentery, and impotency. The gum is astringent, demulcent and tonic. It is used in the treatment of dysentery, haemoptysis in pulmonary tuberculosis, influenza and menorrhagia.

A decoction of the shoots has been used to treat ulcers of the palate, syphilis, leprosy and spider or snake bites.

Butea monosperma (Lam.) Taub.

Family: Fabaceae
Common Name: Palash, Flame of the Forest

- ***Plant Description and Distribution:*** Deciduous trees up to 10 m high, bole crooked, irregular. Bark thick, fibrous, wounds with gum. The tree is leafless while flowering, Flowers large in bunches, bright orange-red. Distributed in Andhra Pradesh, Karnataka, Kerala, Maharashtra, Orissa, Tamil Nadu, West Bengal.
- Medicinal Uses: The fruits and sprouts are used in folk remedies for abdominal tumors in India. The bark is used internally for bleeding. Seeds are used for chest complaints, chronic fevers, earache. In India, the skin ailments; piles and urinary discharges. The juice from the plant as well as the oil is antiseptic. It is an excellent remedy for itch and herpes.The flowers are used for diabetes and biliousness. Roots are used for cleaning gums, teeth, and ulcers.

Eupatorium adenophorum Spreng.

Family: Euphorbiaceae
Common Name: Crofton weed

- ***Plant Description and Distribution:*** It is an evergreen shrub which occurs at any time of the year, at any place. The plant is a common shrub that can be found as weeds, the stem is stout and erect with less branching. The leaves are simple, somewhat blackish green in colour producing black liquid when squeezed. It is about 1 to 2 m in height. It is distributed almost throughout the subtropical and sub-temperate belts of the Himalaya.
- ***Medicinal Use:*** Used for burns, cuts, injuries. It is also used for treatment of wounds, diabetes, blood dysentery, fever, inflammation, jaundice, diarrhea, and stomach pain. It consists of anti-inflammatory, analgesic, antipyretic, antibacterial, antitumor, antifungal, antioxidant, antiseptic and cytotoxic activities.

Ficus Benghalensis L. *var. Krishnae* C.DC

Family: Moraceae
Common Name: Makhan Katori, Sacred Fig Tree

- ***Plant Description and Distribution:*** The Tree height is 10- 15 m. Leaves are cup shaped at the base.Upper surface of leaf is dark green and lower surface is light green in colour. Flower-The taxonomic status of *Ficuskrishnae*is still unclear and currently treated as a variety under *Ficusbenghalensis.* Found distributed in Uttar Pradesh, West Bengal, Karnataka, Kerala, Delhi, Bangalore, Tamil Nadu.
- ***Medicinal Properties:*** It is used in the treatment of ulcers, vomiting, dysentery, fever, inflammations of lever and leprosy.The plant is also used as aphrodisiac, as a tonic in piles and gonorrhea. Stem bark and leaves are useful for diabetes.

Hippophaerhamnoides (L.)

Family: Elaeagnaceae
Common Name: Seabuck thorn

- ***Plant Description and Distribution:*** It is a spiny deciduous shrub with irregular crown in shape with spiny, grey branches with Leaves having linear-lanceolate, alternate, 2-6 cm long, covered on both sides with silvery-white scales, undersides with brown dots. This species is mainly found in Uttarakhand, Himanchal Pradesh, Jammu and Kashmir and Arunachal Pradesh.
- ***Medicinal Uses:*** The seeds and the oils pressed from the pulp both nourish and hydrate the skin. The berries can be ingested or applied topically as an emollient or skin softener. Its leaves and flowers are used in gastrointestinal ulcers, arthritis, gout, and rashes and its fruits are also used to prevent infection and boost immune function.

Murrayakoenigii (L.)

Family: Rutaceae
Common Name: Curry tree, kadipatta

- ***Plant Description and Distribution:*** It is a shrub up to 2.5 m tall and has shiny, dark green, lance-shaped leaflets.The leaves are highly aromatic when rubbed or crushed and are an important flavoring agent and appetizers used in Indian/Asian cuisine. Its stems are dark green to brown, with numerous dots. Distributed along Assam, Karnataka, Maharashtra, Odisha, Punjab, Tamil Nadu, Uttar Pradesh and Indo-Malaya.
- ***Medicinal Uses:*** The bark is helpful in treating snake bites. Its green leaves are used in treating piles, inflammation, itching, fresh cuts, dysentery, bruises, and edema. The roots are purgative to some extent. They are stimulating and used for common body aches. They are used as anthelminthics, analgesics, digestives.

Rhododendron arboreum Sm.

Family: Ericaceae
Common Name: Burans

- ***Plant Description and Distribution:*** It is an evergreen shrub or small tree with a showy display of bright red flowers. up to 10 m high, Flowers, produced in rounded trusses in early spring and may be red, or any shade of pink. It is found in Darjeeling, Himachal Pradesh, Uttarakhand, Arunachal Pradesh and Assam.
- ***Medicinal Uses:*** the leaves are also helpful in the treatment of gout and rheumatism and are used traditionally for healing many diseases like cough, cold, fever, indigestion, headache, etc. Burans possess various therapeutic properties in the treatment of diarrhea, dysentery, irritation, detoxification, bronchitis, fever and asthma.

Santalum album Linn.

Family: Santalaceae
Common Name: Chandan, Sandalwood

- ***Plant Description and Distribution:*** Santalum album is a small tropical evergreen tree attaining a height of 4 to 20 m. The bark of sandalwood trees is dark grey, reddish, dark brown or nearly black. Stem is smooth in young trees, turns rough with deep vertical cracks in mature trees and its wood is hard, very close grained and oily. Sapwood is white, scentless, heartwood is yellowish to brown and strongly scented. Sandalwood is distributed all over the country and more than 90% of it lies in Karnataka and Tamil Nadu covering 8300 sq kms.
- ***Medicinal Uses:*** Sandalwood oil works well as an effective relaxant and sedative which helps to ease muscle spasms. to calm the mind and relieve anxiety, reduce signs of aging and skin inflammation.The oil or the paste can help heal skin diseases such as infectious sores, ulcers, acne and rashes. Sandalwood powder helps smooth and cool the skin, and can be made into a paste, lotion or soap for cleansing, calming and hydrating sensitive or aging skin. Sandalwood acts as a disinfectant, diuretic, expectorant and sedative.

Saracaindica L.

Family: Fabaceae
Common Name: Ashoka

- ***Plant Description and Distribution:*** It is a medium sized tree with a short trunk and spreading canopy, with a height up to 24m.The flowers are orange. It is distributed all over the country particularly in Southern India, Assam, Odisha, West Bengal and Central and Eastern Himalayas up to 750m.
- ***Medicinal properties:*** leaves and flowers have antioxidant properties, aqueous suspension saracaindica flowers are used against ulcer. The tree has been traditionally used in Indian Ayurveda as a uterine tonic and has been indicated in menstrual irregularities. It also stimulates the ovarian tissue.

Taxusbaccata L.

Family: Taxaceae
Common Name: Himalayan Yew, Thuner

- ***Plant Description and Distribution:*** Taxus is a temperate evergreen plant found between 1800 to 3000 m. growing up to 10-20 m tall. The leaves are highly poisonous. The seed cones are highly modified; each cone containing a single seed 4-7 mm long is partly surrounded by a modified scale which develops into a soft, bright red berry-like structure called an aril. In India it is widely distributed in the temperate zone of Himalaya, Himachal Pradesh, Uttarakhand and Arunachal Pradesh.
- ***Medicinal Uses:*** The important anti-cancer drug, called taxol, is extracted from the plant leaves. The paste prepared from its bark is applied as a plaster on fractured bone. Extract from the bark and leaves is also used for the treatment of various diseases like bronchitis, asthma, poisonous insect bites and also as an aphrodisiac.

Terminalia chebula Retz.

Family: Combretaceae
Common Name: Harad, Haritaki

- ***Plant Description and Distribution:*** It is found in the Sub Himalayan region to West Bengal and Assam, ascending up to the altitude of 1500 m in the Himalayas. This tree is wild in forests of Northern India, central provinces and Bengal, common in Madras, Mysore.
- ***Medicinal Uses:*** The fruit is useful in ophthalmia, hemorrhoids, dental caries, bleeding gums, and ulcerated oral cavity. Its paste with water is found to be a healing capacity for wounds. Its decoction is used as a gargle in oral ulcers, sore throat. It is good to increase appetite, digestive aid, liver stimulant, stomachic, gastrointestinal prokinetic agent, and mild laxative. It is used in nervous weakness, nervous irritability.

Thuja occidentalis L.

Family: Cupressaceae
Common Name: Thuja, Morpankhi, White Cedar, Arbovitae

- ***Plant Description and Distribution:*** Thuja is an evergreen coniferous tree up to 15m. The bark is red-brown, furrowed, and has fan-like branches and scaly leaves. *T. occidentalis* cones are ellipsoidal. It is commonly distributed in Uttarakhand, Himachal Pradesh, Arunachal Pradesh, Assam and Jammu and Kashmir.
- ***Medicinal Use:*** It is commonly used to treat liver diseases, bullous bronchitis, psoriasis, enuresis, amenorrhea, cystitis, uterine carcinomas, diarrhea, and rheumatism. Thuja is also used for respiratory tract infections such as bronchitis, bacterial skin infections, and cold sores.

FUTURE PERSPECTIVE

Herbal and traditional medicine is considered as a promising future medicine for the management of health care. In recent years there has been a swing in the universal trend from synthetic to herbal medicine.In the beginning, the trial and error method was used to treat illnesses or even simply to feel better, and in this way, to distinguish useful plants with beneficial effects (Kunle et al., 2012).The unhealthy lifestyle and environmental toxins increase the risk of diseases. The side effects, overuse and misuse of allopathic drugs are also a major concern. The scientific progress in allopathic medicine and modern facilities also resists the growth of traditional medicine. Still, about 70% rural populations of India are believed in traditional medicine for primary healthcare (Mafuva et al., 2014; Pandey et al., 2013).The advancement in medical science has also led to the increased efficacy in pharmacological effects of medicinal plants.

WHO in 2013 developed 'WHO Traditional Medicine Strategy 2014–2023' to integrate traditional and complementary medicine to promote universal healthcare and to ensure the quality, safety and effectiveness of such medicines. The organization has stated that the trade in medicinal plants, raw materials, and drugs is growing at an annual growth rate of approximately 15% (New Look to Phytomedicine, 2019). Most countries are endowed with vast resources of medicinal and aromatic plants. Resurgence of interest in green products globally has created an expanding market for plant based products that would be increasing in future. Promoting ethnobotanical studies and research will help in the discovery of crude drugs. Medicinal plants have a hopeful future, as the phytochemical composition and the potential health benefits of many species have not yet been studied or still need to be more deeply investigated (Kia et al., 2018). India has one of the richest plants based medical traditions in the world. There are estimated to be around 25,000 effective plant-based formulations, used in folk medicine and known to rural communities in India. There are over 1.5 million practitioners of the traditional medicinal system using medicinal plants in preventive, promotional and curative applications. It is estimated that there are over 7800 medicinal drug-manufacturing units in India, which consume about 2000 tons of herbs annually (Sheetal et al., 2008). The herbal medicine is still the most preferred method of treatment and provides the safety and effectiveness of a therapeutic product. Educating the young generation and students about the importance and use of such medicinal plants might help them to develop understanding of natural remedies and its economic aspects.

REFERENCES

Baul TK and Mahiuddin M. (2011).Certain medicinal tree species in the forests of Bangladesh. *Journal of Tropical Medicinal Plants*, 12(2): 225-228.

BGCI (2021). State of the World's Trees. BGCI, Richmond, UK.

Census of India (1961). District Census Handbook, Mayurbhanj.Superintendent. Orissa, Government Press, Cuttack.

Global Initiative for Traditional Systems (GIFTS) of health food and agriculture organization of the United Nations, Rome 1997, Non Wood Forest Products 11, Medicinal plants for forest conservation and health care.

Jamshidi-Kia. F., Lorigooini. Z., Amini-Khoei, H. (2018). Medicinal plants: Past history and future perspective. *J. HerbmedPharmacol.*, 7, 1–7.

Kewessa G., T. Abebe, A. (2015). Demessie Indigenous knowledge on the use and management of medicinal trees and shrubs in Dale District, Sidama Zone, Southern Ethiopia Ethnobot. *Res. Appl.*, 14 , pp. 171-182.

Kunle O.F., Egharevba H.O., Ahmadu P.O. (2012). Standardization of herbal medicines-A review. *Int. J. Biodivers. Conserv.* 4:101–112. doi: 10.5897/IJBC11.163.

Lewington, A. (1993). Medicinal Plant and Plant Extracts: A Review of Their Importation into Europe. Cambridge, UK; Traffic International.

Newton AC. (2008). Fauna & Flora International, *Oryx*. 42(2): 195-205.

Rakhee Dimri, Prapti Kondaji, Sanjeet Kumar and Sweta Mishra (2021). book Medico-Biowealth of India, Vol III.

Schippmann U, Leaman JD and Cunningham AB. (2002). Impact of cultivation and gathering of medicinal plants on biodiversity: Globaltrends and issues, Inter-Departmental Working Group on Biological Diversity for Food and Agriculture, Rome.

Mafuva C., Marima-Matarira H.T. (2014). Towards professionalization of traditional medicine in Zimbabwe: A comparative analysis to the South African policy on traditional medicine and the Indian Ayurvedic system. *Int J Herb Med.* 2: 154–161.

New Look to Phytomedicine Advancements in Herbal Products as Novel Drug Leads 2019, 3-13.

Pandey M.M., Rastogi S., Rawat A.K.S. (2013). Indian traditional Ayurvedic system of medicine and nutritional supplementation. *Evid Based Complement Altern Med.*, 1–12.

Sheetal Verma and S.P. Singh (2008). Current and future status of herbal medicines, *Veterinary World*, Vol. 1(11): 347-350.

CHAPTER 27

Ethical Use of Technology and Digital Citizenship: Concerning NEP 2020 and Atamnirbhar Bharat

Veena Singh

PhD Scholar, IASE, Jamia Milia Islamia, New Delhi, India

Abstract

Atmanirbhar Bharat and NEP 2020 are the two initiatives launched by the government of India to become more independent and to reform the education system. NEP 2020 has recommended the use of technology in education. Though the introduction of technology into the educational system has complicated ethical issues and classroom ethics, students are taught ethics from the start in all schools these days. Teachers instruct them about the rules and regulations of the previous school where they attended, and they learn not to cheat or hurt anyone. Digital citizenship fosters ethical technology use, helps students use technology safely and respectfully, and helps them understand how using technology affects them, others, and society. Additionally, it raises awareness of internet dangers like scams, identity theft, and cyberbullying. In the current digital era, digital citizenship is crucial because it fosters responsible technology use, critical thinking on online safety, the positive effects of technology, and upholding ethical standards when utilizing it. The focus of the study is to examine the significance of digital citizenship, the ethical use of technology in education, and raising awareness of these topics among students and in the education system.

Keywords: *Atmanirbhar Bharat, Technology, Ethics, Digital citizenship, NEP 2020*

INTRODUCTION

Atmanirbhar Bharat Abhiyan and the New Education Policy NEP 2020 are the two main initiatives launched by the government of India. Recently, the Atmanirbhar Bharat Abhiyan was introduced on May 2020 to encourage economic independence and lessen India's reliance on paid imports from abroad. 2020 was introduced in July 2020 as part of the national education policy to reform the Indian educational system. The NEP 2020 emphasizes the liberating power of digital technology to enhance access, equity, and equality in education across all levels. This includes making provisions for the development of digital infrastructure, such as high-speed Internet connectivity and digital devices, to ensure that all students, regardless of location or socioeconomic background, can learn. 2020 digital transformation plays an important role in reshaping the educational landscape of India. can benefit from digital learning resources and tools additionally the NEP 2020 promotes the integration of technology into the teaching and learning process facilitating personalized learning experiences interactive pedagogies and adaptive assessment practices furthermore the policy encourages the use of digital platforms for teacher professionals' development of administrative task educational research for stream culture of innovation and collaboration within the education ecosystem overall digital transformation under NEP2020 is Invision as a catalyst for democratizing education in powering learners and educators and driving sustainable progress towards educational excellence in India.NEP 2020 places a high priority on the development of 21st-century skills, which include digital literacy, critical thinking, creativity, communication, and collaboration. These skills are vital in preparing students for the challenges of the modern world. It is crucial to teach students about the ethical use of technology as part of their preparation for global citizenship, as this will help them understand digital citizenship. Students who use technology ethically may consider how to care for it and investigate safe and efficient ways to use it for research and education. Students will learn about the protections that come with using technology as well as how to keep others and themselves safe from cybermedias. They'll comprehend how to produce content ethically and responsibly.

DIGITAL CITIZENSHIP

Digital citizenship encompasses a variety of behaviors, such as online safety, and is defined as the appropriate use of technology and online resources, such as social media messages and other digital tools. Utilizing technology in an ethical manner and communicating digitally In a world where technology is developing quickly and being used more and more in daily life, digital citizenship is crucial because it gives structure to tasks, teaches people how to use technology responsibly, and helps them interact with others. It encompasses a variety of knowledge, skills, and mindsets related to using technology, including digital literacy, ethics, and etiquette that outline safety and responsibility. The first aspect of student digital citizenship is knowing and abiding by appropriate and practical rules of conduct when utilizing technology.

Why an Understanding of Digital Citizenship is Important for Students

Given that NEP 2020 advocates for the use of technology in the classroom, it is critical that students learn about digital citizenship. Comprehending digital citizenship facilitates pupils' ethical utilization of technology. Teaching children about digital citizenship is crucial because they will learn how to use technology responsibly, effectively, and without hurting themselves or others. They will also become more responsible users of technology. Understanding offers a variety of behaviors, such as ethical behavior, responsible digital communication, and online safety. It offers a structure for the issues that must be resolved and offers a framework for learning how to use technology appropriately. The ethical consequences of using increasingly complex technology must be taken into account, especially when it comes to data privacy, security, and equity. Users need to use caution when evaluating the veracity of the material offered and refrain from if anything produced on these sites is legitimate. Everyone must be trained to recognize and capitalize on the possibilities of technology, and those who use these tools must strike a balance between their rights to use them and their obligations to share the source of the information.

Components of Digital Citizenship in Education

Practicing safe use of technology including creating strong passwords securing personal information and understanding the threats posed by viruses phishing and other cyber threats. The following are the elements of digital citizenship in education:

1. **Digital literacy**, which includes critical thinking about the validity and dependability of online information, as well as the ability to use technology to find news and create information efficiently in a digital environment.
2. **Digital etiquette:** knowing and adhering to appropriate online conduct, such as showing respect to others, abstaining from cyberbullying, and acknowledging the impact of one's actions on others.
3. **Digital ethics:** knowing the rules both legally and morally, such as those pertaining to copyright, digital rights, and responsible use of resources and software.
4. **Digital responsibilities:** appreciating the well-balanced rights to information access, freedom of speech, and privacy while also being aware of the obligations that go along with those rights, such as upholding them.
5. **Digital wellness:** learning about the psychological and physical effects of using digital technology, as well as creating safe and healthy behaviors
6. **Online safety:** using technology responsibly, which includes making strong passwords, protecting personal data, and being aware of the risks associated with viruses, phishing scams, and other online dangers.

Importance of Digital Citizenship for Students

For students' digital citizenship is important for different reasons:

- **Safety:** Understanding how to interact safely online reduces the risk of encountering dangerous situations or engaging in risky behaviour.
- **Integrity:** developing a sense of digital ethics helps students understand the importance of honesty and integrity in their academics.

- **Future-readiness:** as higher education and workplaces increasingly rely on digital technology students need to be proficient and ethical digital citizens to prepare them for future education and professional pursuits.
- **Participation:** in a world where many societal issues are discussed and acted upon online being a good digital citizen means being able to engage effectively and responsibly in the community.

Challenges and Solutions in Digital Citizenship Education

- **Quick Technical Developments:** Teachers find it difficult to stay current on the newest techniques, resources, and possible hazards because of the ever-changing digital environment.
- **Differential Student Exposure and Access:** Students have different histories and levels of access to and experience with technology, which can lead to differences in comprehension and participation.
- **Encouraging Students to Use Digital Tools Responsibly and Independently:** It can be challenging to strike the correct balance between ensuring students' safety online and giving them the freedom to do so.
- **Including Diverse Learners:** It might be difficult to meet the needs of all pupils while also making digital citizenship engaging and relevant for all learning types.
- **Parental Involvement:** Education regarding digital citizenship in schools frequently falls short of what students are taught at home.

Implication of Digital Citizenship in Education

Educating students about digital citizenship requires the integration of an appropriate and age-appropriate digital curriculum. This requires collaboration between teachers, parents, and students, supported by school policies that promote the safe and responsible use of technology. Effective digital education should not only include instructions on how to be a good citizen but also encourage students' responsibility. Communication with technology includes understanding the cultural and social impacts of digital technology through voting and promoting resources for student empowerment.

Methodology

Methodology Secondary data has been used for the preparation of this article. The NEP 2020 document has been used for understanding and analyzing the same. Apart from these, data from online sources and various reports relating to ethics and digital citizenship and NEP 2020 have been used for the preparation of the article. This article purely reflects the sole opinion of the authors based on their understanding of NEP 2020 and ethics and digital citizenship.

ANALYSIS AND CONCLUSION

Due to the rapid and constant change in the digital world, which necessitates learning new skills, assessments of various social groups' skills must also take into account the quality of education that digital knowledge must impart to them, as well as their access to resources via the Internet and technology It is more crucial than ever for students to learn digital citizenship and ethical technology use in today's increasingly digital society. Digital citizenship involves using technology in a responsible, ethical, and productive way by developing and utilizing successful digital citizens. Teachers can equip pupils with skills that will empower them. Research on digitization and its effects on society, including how it affects modern people's expectations, attitudes, abilities, and evaluations, must be conducted in this context.

REFERENCES

Agamba, J. J., & Keengwe, J. (2012). Pre-Service Teachers' Perceptions of Information Assurance and Cyber Security. *International Journal of Information and Communication Technology Education* (*IJICTE*), 8 (2), 94-101.

Al-Zahrani, A. (2015). Toward digital citizenship: Examining factors affecting participation and involvement in the Internet society among higher education students. *International Education Studies*, 8(12), 203.

Elmali, F (2020). A study in digital citizenship: Preschool teacher candidates vs. computer education & instructional technology teacher candidates. *Turkish Online Journal of Distance Education*, vol. 10, 251-269.

Milenkova,V (2021). Digital citizenship and digital literacy in the conditions of social crisis. *Interactive technology & smart education*, vol. 10(4),40.

Pangrazio,L (2021). Digital rights, digital citizenship, and digital literacy: what's the difference? *Approaches in educational research*, vol. 10, 2254-7339.

CHAPTER 28

Empowering Atmanirbhar Bharat: The Role of Technology and Innovation in Entrepreneurship and Job Creation under National Education Policy

Ansh Verma

Scholar, Himalayan Institute of Technology, Dehradun, Uttarakhand, India

Abstract

The National Education Policy (NEP) 2020 represents a significant shift in India's education system, bringing with Atmanirbhar Bharat's vision of encouraging entrepreneurship and job creation. This paper explores the critical role of technology and innovation in promoting entrepreneurship and job creation within the context of the NEP 2020. By adding technological advancements and innovative methodologies into the curriculum, NEP 2020 aims to prepare students for entrepreneurial endeavors and rising job markets. The policy's emphasis is on experiential learning, critical thinking, and vocational training which aims to fill the gap between academic knowledge and practical application. The paper findings are that the strategic implementation of NEP 2020, with a strong focus on technology and innovation, can make a significant contribution to the realization of Atmanirbhar Bharat by developing a generation of skilled entrepreneurs and workforce professionals.

Keywords: *Atmanirbhar Bharat, Entrepreneurship, NEP 2020, Technology, Innovation*

INTRODUCTION

The National Education Policy (NEP) 2020 aims to modernize India's educational system to achieve Atmanirbhar Bharat, the country's ideal of self-sufficiency. Major changes are made to this approach to better match academic and vocational programs to the changing needs of the local and global markets. By highlighting the integration of technology and innovation, NEP 2020 aims to fill the skills gap between traditional education and the practical skills needed in the modern workforce by promoting experiential learning, critical thinking, and vocational training. This will encourage entrepreneurship and job creation. One of the most vital aspects of NEP 2020 is its focus on developing a strong digital infrastructure and adopting educational technology. The policy seeks to encourage creative minds and problem solvers by supporting an integrated approach to development. To ensure that students have the ability to navigate and participate in the digital economy, this is achieved by integrating digital tools and resources into the classroom environment.

The National Education Policy 2020 also highlights the value of collaborations between industry and academics. The policy intends to establish an environment that supports both economic growth and ongoing improvement of education by promoting collaboration between educational institutions and industries. These collaborations play an essential part in bridging the knowledge gap between theory and practice by giving students practical knowledge that applies to real-world challenges. The policy also intends to democratize education by making it available to all members of society, no matter their socioeconomic status. This includes initiatives that promise that each child has an opportunity to be successful by offering high-quality education in rural and remote areas. NEP 2020 wants to do this to develop a fairer and accessible educational environment.

Also, by supporting the use of regional languages in the educational system, the policy supports diversity and the preservation of India's rich cultural legacy. This enhances both understanding and memory retention by allowing students to study in the language in which they are most comfortable, while also contributing to the preservation of diversity in languages. NEP 2020 aims to establish a nation that is self-sufficient and continues its growth by focusing on the integration of innovation and technology and industry-academia cooperation. With this approach, the policy aims to equip students with the skills necessary to succeed in a constantly evolving global environment, allowing them to contribute to it.

LITERATURE REVIEW

NEP 2020 and Educational Reforms

The National Education Policy (NEP) 2020 represents a major change of direction in India's education environment, as noted by recent analysis by Agarwal and Rajkumar (2021) and the Ministry of Education, Government of India (2020). The goal of this policy is to create a comprehensive, adaptable, and multidisciplinary educational system that meets the changing demands of the present day. NEP 2020 recognizes each student's unique talents and potential and tries to develop them accordingly. Highlighting experiential learning, critical thinking, and vocational training—all of which are believed to be crucial for bridging the gap between academic knowledge and practical applications. NEP 2020 aims to transform traditional learning techniques used in Indian education and introduce a more dynamic and engaging learning environment by giving priority to these aspects. The integration of experiential learning and vocational training not only improves students' employability but also gives them the tools they need to deal with issues that occur in the real world. The educational goals of NEP 2020 are in keeping with current worldwide trends, which place a strong importance on lifelong learning and complete development.

Technology and Innovation in Education

The integration of technology into education is a top priority of the National Education Policy 2020 (NEP). According to Sharma (2020), information and communication technology (ICT) plays an important role in increasing access to education and entrepreneurial opportunities in underdeveloped countries. This point of view is supported by NEP 2020, which aims to improve education through digital platforms in terms of efficiency, inclusivity, and accessibility. Similarly, Drucker (1985) highlights the role that innovation and entrepreneurship play in promoting growth in the economy. He is in support of including entrepreneurship education in the curriculum to help students get ready for the demands of a changing global economy. This idea is supported by NEP 2020, which promotes an entrepreneurial mindset and skill set among learners through new teaching techniques and digital resources. The integration of technology, innovation, and entrepreneurship in education not only prepares students for success in the digital age but also accelerates socioeconomic progress. NEP 2020's dedication to digitalization and entrepreneurship shows a forward-thinking approach to education, attempting to reduce the digital gap, promote inclusive growth, and create a fairer society.

Digital Transformation in Education

The proper implementation of NEP 2020 requires the integration of digital technologies into the educational process. The UNESCO research report on digital transformation in education (2020) provides valuable insights into the most effective ways to use technology to improve student performance. It highlights how crucial it is to use innovative teaching and learning strategies including virtual classrooms, interactive multimedia resources, and online platforms. These resources support individualized learning experiences that are personalized to the needs of each student in addition to making high-quality education more accessible. Also, The NITI Aayog's Atmanirbhar Bharat (2020) policies highlight the importance of a strong digital infrastructure in supporting educational programs and boosting entrepreneurial activities. Learning environments that are engaging and enable students to develop critical thinking, problem-solving, and digital literacy skills—all necessary for success in the professional world of the twenty-first century—can be established by educational institutions through investments in digital technologies. Digital platforms also provide chances for collaboration, knowledge sharing, and skill development, all of which support learners' overall growth. In summary, the implementation of educational technologies is critical to achieving NEP 2020 goals and advancing the vision of Atmanirbhar Bharat. By adopting digital transformation in education, India can create a workforce that is ready for the future and has the knowledge and abilities required for survival in a quickly changing global market.

Impact of NEP 2020 on Employability

Recent research examined the impact of the National Education Policy of 2020 on employability, highlighting the policy's priority for skill development and vocational training. Manimala (2021) explores the entrepreneurial sides of NEP, underlining the program's focus on encouraging students' creativity and innovation. In the same way, Anjali and Ramesh (2021) highlight how the strategy integrates vocational training and practical training into regular education intending to improve graduates' preparation for the workforce. In addition, the World Economic Forum's 2020 Future of Jobs Report highlights the growing need in today's workforce for abilities like as problem-solving, critical thinking, and technological competence. NEP 2020 is in alignment with these findings as it places a high priority on students developing these kinds of abilities, which guarantees their applicability and competitiveness in the changing job market. Through the establishment of a link between educational outcomes and market demands, NEP 2020 targets to produce graduates with both academic ability and the practical skills necessary for success in the workplace. This focus on employability is an indication of a strategic approach to education that adapts to the changing demands of the international market. In conclusion, it reviewed the highlights of how NEP 2020›s priority for skill development, vocational training, and alignment with the changing demands of the job market enhances employability. Through the provision of necessary abilities and skills, the policy aims to improve graduates› employment opportunities and promote general social and economic growth.

Industry Academia Synergy

Recent research has highlighted the vital part that industry-academia partnership plays in promoting innovation and employment growth. To create an innovative environment, strong industry-academia links are important, as highlighted in the GEM India Report (2021). The revolutionary effect of these collaborations or partnerships on the quality and applicability of education is further shown by studies from Balaji and Ramaswamy's research on digital platforms and entrepreneurship. By providing students with practical experience and exposure to real-world industrial practices, these partnerships improve their employability and prepare them for the workforce. Through creating stronger connections between academic institutions and industries, NEP 2020 aims to develop an education system that is more dynamic and adaptable, keeping up with the changing demands of the industries. NEP 2020›s strategic goal for promoting innovation, entrepreneurship, and sustainable employment growth is reflected in this emphasis on industry-academia synergy.

National Apprenticeship Promotion Scheme (NAPS) by The Ministry of Skill Development and Entrepreneurship

The goal of Ministry of Skill Development and Entrepreneurship's (MSDE) National Apprenticeship Promotion Scheme (NAPS) is to close the divide between education and industry requirements. NAPS promotes the involvement of businesses in training and improving the skills of individuals, increasing their chances of getting hired. Kumar and Singh (2019) NAPSinitiated apprenticeship programs significantly improve technical skills and employment opportunities, especially in the manufacturing sector.Similarly, apprentices reported high levels of satisfaction with the quality oftrainingand job opportunities(Sharma and Gupta,2020).NAPS also aims to promote inclusivity by encouraging marginalized groups toparticipate.However, barriers such as insufficient knowledge among companies and candidates continueto exist, requiring focused campaigns for outreach. NAPS plays a vital role in closing the skills gap and boosting job opportunities.

Government Initiatives and Growing Investment for Job Creation

Chief Economic Advisor (CEA) v. Anantha Nageswaran expressed a positive view on employment expansion, attributing it to effective government interventions and better financial situations. Important factors suggest a change in previous patterns, as non-food credit growth approaches 20% and companies and banks show improved balance sheets. Data from 2021–22 showed a decrease of 15 lakh jobs in agriculture, with manufacturing and services each experiencing a growth of 37 lakh jobs, and construction creating 19 lakh jobs. This change signals strong expansion in the manufacturing and construction sectors. The Government is working on skill development programs, contributing to the Employees' Provident Fund Organization (EPFO), implementing the New Education Policy, and carrying out structural human development reforms. Investing in physical infrastructure is important for supporting the expansion of industry and manufacturing, resulting in jobs. Additionally, the improvement in financial institutions' health serves to reinforce the good trend even more. However, it is recognized to make this expansion easier, simpler regulatory and tax policy adjustments are required. This literature highlights how strategic initiatives and investments work together to promote job creation in India.

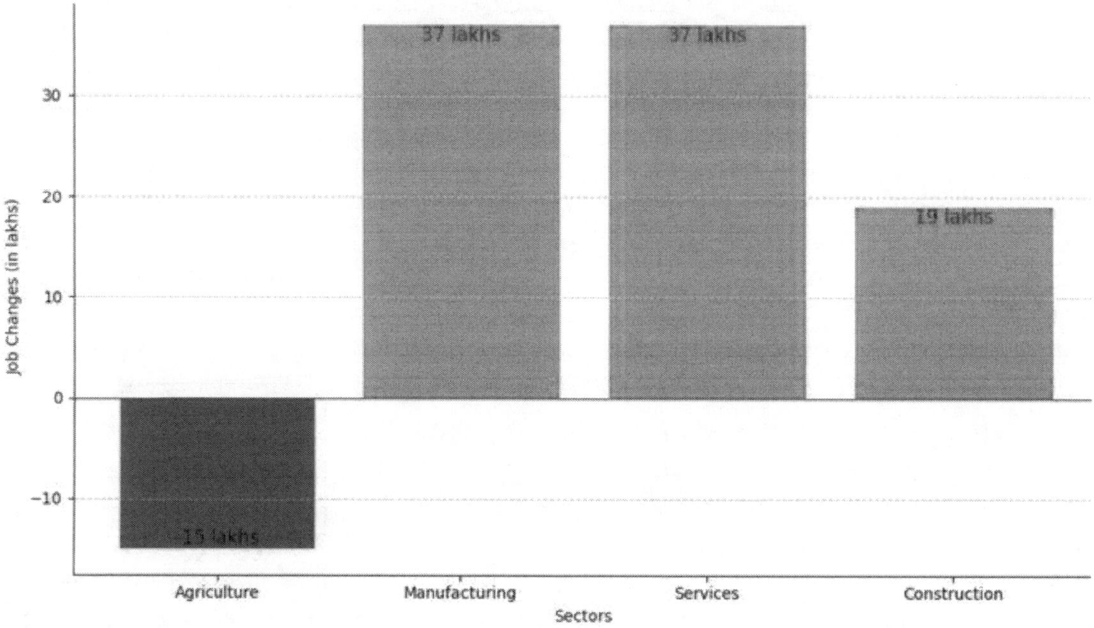

Figure 1: Job Changes by Sector (2021–22)

Innovation in Governance

BPAC's study on "Innovation in Governance" demonstrates how technological projects like the Smart Cities Mission and the Digital India Project are crucial to improving public service delivery and transparency. These initiatives point out the possible impacts of NEP 2020 on job creation and company development while supporting the goals of a self-reliant India. The BPAC observations highlight the importance of utilizing innovation and technology to drive economic growth and address current governance challenges. These initiatives contribute to the development of an atmosphere that is favorable for economic expansion and the creation of job opportunities by supporting an effective and transparent governance system.

Government Initiatives for Startup Growth

India's successful startup environment, backed by strong government efforts, has established the nation as a leading center for innovative businesses. With 1.2 million workers and over 123,900 recognized startups as of March 2024, and is at the forefront of innovation and technology globally. Positive policies, tax benefits, streamlined business procedures, and regulatory changes have built a climate that supports the growth of new businesses. Initiatives such as the Startup

India Seed Fund Scheme and technical assistance provided by TIDE 2.0 are being to strengthen this environment. The Atal Innovation Mission and the Startup India Initiative show the government's dedication to promoting entrepreneurship.

India's startup ecosystem is set for rapid expansion with 114 unicorns worth $350 billion, supporting the objectives of Atmanirbhar Bharat and job creation as per NEP 2020. If India continues to support and maintain this vibrant ecosystem, it could benefit from increased economic growth and employment opportunities.

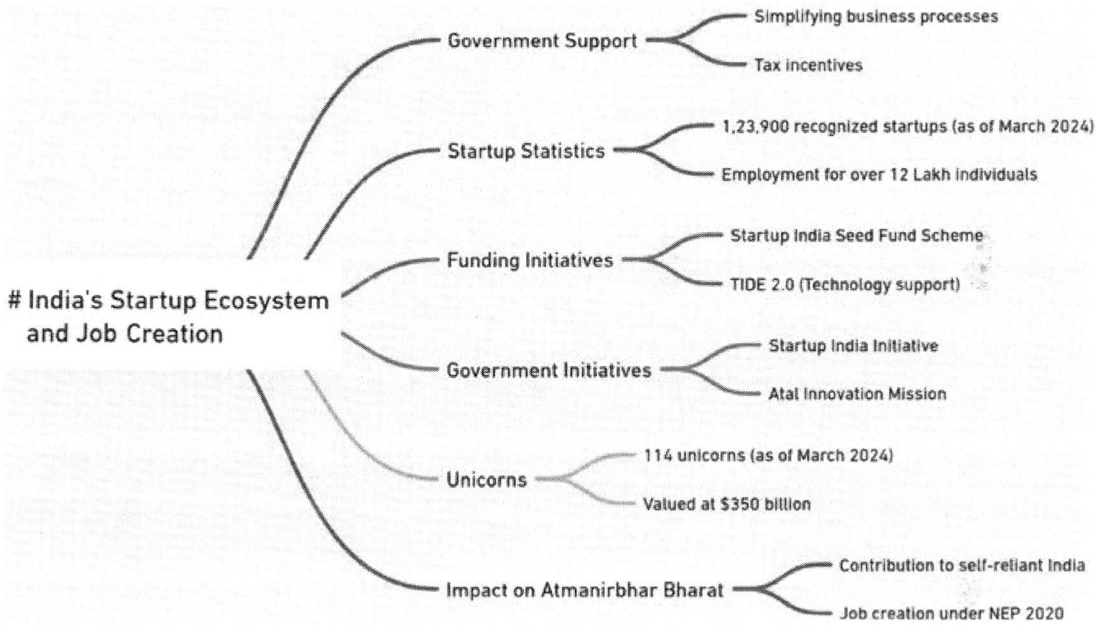

Figure 2: India's Startup Ecosystem and Job Creation

CONCLUSION

The NEP 2020 brings about a big change in India's education system, in line with Atmanirbhar Bharat's vision, by encouraging entrepreneurship, job creation, technology, and innovation. NEP 2020 combines technological advancements and innovative methods in the curriculum to connect academic learning with real-world practice, equipping students for today's workforce and entrepreneurial endeavors. The policy focuses on hands-on learning, analytical thinking, and practical training to provide students with essential skills. Digital transformation is essential, improving access, diversity, and customized learning opportunities. Collaborations between industry and academia continue to drive innovation and economic growth, as government programs like the National Apprenticeship Promotion Scheme (NAPS) also contribute to job growth and skill enhancement. Initiatives like the Smart Cities Mission and Digital India improve public service delivery and transparency, in line with the objectives of NEP 2020. Government policies and funding support a flourishing startup ecosystem, highlighting the dedication to promoting a self-sufficient and economically strong India.

Overall, NEP 2020's emphasis on technology, innovation, and collaboration between industry and academics sets the path for India to develop skilled entrepreneurs and professionals. An effective execution, continuous assessment, and collaboration across all aspects are essential to fully utilize the potential of NEP 2020, enabling India to create a skilled workforce prepared for a rapidly evolving global market.

REFERENCES

Anjali, S., & Ramesh, R. (2021). Impact of NEP 2020 on Higher Education and Employability. *Journal of Educational Research and Innovation.*

Agarwal, A.K., & Rajkumar, R.(2021). India's National Education Policy 2020: Implementation Challenges and Opportunities.

Balaji, N., & Ramaswamy, S. (2020). The Role of Digital Platforms in Fostering Entrepreneurship: An Analysis. *International Journal of Business Innovation and Research.*

BPAC (n.d.). Innovation in Governance. https://bpac.in/innovation-in-governance/

Drucker, P.F. (1985). Innovation and Entrepreneurship: Practice and Principles.

Economic Times. Government initiatives, growing investment to create more job opportunities during decade: CEA.

Global Entrepreneurship Monitor (GEM) India Report (2021).

InvestIndia (n.d.). The Role of Government Initiatives in Boosting Startups. https://www.investindia.gov.in/team-india-blogs/role-government-initiatives-boosting-startups

Kumar, A., & Singh, R. (2019). Impact Assessment of National Apprenticeship Promotion Scheme (NAPS) on Manufacturing Sector.

Manimala, S. (2021). Entrepreneurship and Innovation: A Study of India's New Education Policy. *Journal of Entrepreneurship Education.*

Ministry of Education, Government of India. NEP 2020: Transformative Reforms in Education.

Ministry of Human Resource Development, Government of India. National Education Policy 2020.

NITI Aayog. Atmanirbhar Bharat Abhiyan: Strategies for Self-Reliant India.

Sharma, S.D. (2020). Technology and Innovation in Entrepreneurship: The Role of ICT in Developing Countries. *Journal of Entrepreneurship and Innovation in Emerging Economies.*

World Economic Forum (2020). Future of Jobs Report.

UNESCO (2020). Digital Transformation in Education.

CHAPTER 29

A Critical Study on the Role of Government Scheme for Entrepreneurship Development

Dr. Jai Jayant[*] & Neha Thapliyal[**]

[*]Assistant Professor, [**]Research Scholar
School of Business Studies, Jigyasa University, Dehradun, India

Abstract

Governments worldwide implement various initiatives to foster entrepreneurship and support small business development programmes. Emerging as the world's third-largest startup ecosystem. Government policies and programs promote entrepreneurship and investing in new ventures. This paper examines various government policies and programs towards the development of entrepreneurship in INDIA. The Government of India and the Ministry of Micro, Small, and Medium Enterprises have launched more than 21 schemes to empower startups and MSMEs in India. In this paper, we have listed out 21 govt. schemes to promote entrepreneurship in India that are specially for startups or entrepreneurs who are looking to get help from these schemes such as Startup Leadership Program (SLP), Startup India Initiative and Credit Guarantee Trust Fund, etc. The government has prioritized the establishment of high-quality incubators and accelerators across the country to foster collaboration, innovation, and knowledge exchange among entrepreneurs. The research underscores the vital role played by government funds and initiatives in overcoming financial barriers and empowering startups to pursue innovative ideas. As of March 14,2024, there were 1,23,900 DPIIT -recognised startups, with at least one recognised startup in every state as of December 31, 2023. The recognised startups have generated direct employment for over 12 lakh individuals. Startups have been at the forefront of innovation -introducing new technology, product and services that could change the world while generating jobs, establishing valuable enterprises' & innovations.

Keywords: *Entrepreneurship, Startup Schemes, Employment, Innovations*

INTRODUCTION

Entrepreneurship development is one of the mechanisms adopted by the Government of India towards the creation of job opportunities. Governments spend a lot of money to promote entrepreneurship and the growth of small business development programmes. In order to ensure regional development, they also support business in rural and underdeveloped area. They launched a number of programs & policies to assist entrepreneurship with marketing, finance, technique and skill development in order to help them accelerate and adapt to changing business trends. Entrepreneurship has long been recognized as a fundamental driver of economic growth and development in nation. In this research paper we analyse various government schemes that encourage various aspects of entrepreneurship, including the creation of new businesses, the introduction of innovative products and services, and the impact on job creation, to assess its unique contributions to economic development. As per records, the scale and magnitude of startups in India are the highest they've ever been, making the country one of the fastest-growing startup ecosystems in the world. Government schemes such as Startup India, Standup

India, and Atal Innovation Mission (AIM) have been instrumental in facilitating the growth of startups in India. These schemes provide access to funding, tax incentives, mentorship, and incubation support, allowing entrepreneurs to overcome initial challenges and establish sustainable businesses. Governments started to implement general and specific policies that promote entrepreneurial activities.

Aatmanirbhar Bharat App Innovation Challenge

On 4th July 2020 The scheme launched named "Aatmanirbhar Bharat App Innovation Challenge", which aim to provide financial assistance, mentorship, and support to the budding entrepreneurs of the country. It include the Startup India Scheme, Standup India Scheme. The startups of India to come together and develop 'Made in India' applications. All the tech and startup community to develop world-class mobile applications. The challenge was launched under the Digital India mission, with The Atal Innovation Mission (AIM) which is an endeavour of Niti Aayog. The Government of India banned 59 Chinese mobile applications such as TikTok, ShareIt, Xender, etc. from the Indian market.

Samridh

On 25th August 2021, Minister of Electronics Information and Technology (MeitY) launched the SAMRIDH scheme, which stands for Startup Accelerators of MeitY for Product Innovation, Development, and growth, The scheme will support to select and accelerate potential IT based startups. It will provide first round of funding of up to ₹40 lakh in the business of incubation for more than 3 years and supported more than 50 start-ups of which at least 10 have received non-public investment OR having targeted accelerator programs with experience of running at least 3 cohorts with activities listed as desirable under SAMRIDH.

Startup India Seed Fund

On 16th January 2021, the scheme has launched of the 'Startup India Seed Fund' to help startups and support ideas from aspiring entrepreneurs. An Experts Advisory Committee (EAC) has been constituted by Department for Promotion of Industry and Internal Trade (DPIIT), which will be responsible for the overall execution and monitoring of this Scheme. DPIIT has created Startup India Seed Fund Scheme (SISFS) with an outlay of INR 945 Crore and will support an estimated 3,600 entrepreneurs through 300 incubators in the next 4 years (2021-2025). Preference would be given to startups creating innovative solutions in sectors such as social impact, waste management, water management, financial inclusion, education, agriculture, food processing, biotechnology, healthcare, energy, mobility, defence, space, railways, oil and gas, textiles, etc.

Startup India Initiative

On 16th January 2016, under the Startup India initiative, eligible companies can get recognised as Startups by DPIIT, in order to access a host of tax benefits, easier compliance, IPR fast-tracking & more. Over 96,000 startups are registered with the DPIIT till March 2023. Turnover of the entity for any of the financial years since incorporation/ registration has not exceeded one hundred crore rupees. It is incorporated as a private limited company (as defined in the Companies Act,

2013) or registered as a partnership firm (registered under section 59 of the Partnership Act, 1932) or a limited liability partnership (under the Limited Liability Partnership Act, 2008) in India.

Startup Leadership Program

It has launched in 2016 in India, SLP is designed as a "highly selective 6-month world-class training program and lifetime network". This startup initiative first began in Boston in 2006, and has spread its wings in over 14 countries and 28+ cities, and has helped 2000+ startups, influenced 3600+ entrepreneurs. In addition to programs like SLP, there are also various government initiatives that offer help for business start-ups.

Aspire

A Scheme for Promotion of Innovation, Rural Industries and Entrepreneurship (ASPIRE) is a Government of India initiative, promoted by the Ministry of Micro, Small and Medium Enterprises (MSME).The mentioned scheme was launched in 2015 The scheme has aimed at increasing employment, reducing poverty, and encouraging innovation in rural India. The total budget of the scheme initially was INR 200 crores for the period of 2014 2016.

Pradhan Mantri Mudra Yojana (PMMY)

Micro Units Development Refinance Agency (MUDRA) The government has introduced this scheme to support small businesses in India. In April 8 2015, the government allocated INR 10,000 crores to promote startup culture in the country. The MUDRA banks provide startup loans of up to INR 10 lakhs to small enterprises, and businesses, which are non-corporate, and non-farm small/micro-enterprises. MUDRA comes under Pradhan Mantri Mudra Yojana (PMMY) which was launched on 8 April 2015.

The loans have been categorized as: 1. Shishu (up to ₹50,000), 2. Kishore (between ₹50,000 and ₹5 lakh) and 3. Tarun (₹10 lakh).

As of March 24, 2023, about ₹23.2 lakh crore has been sanctioned in 40.82 crore loan accounts, Any Indian citizen with a business plan for a non-farm income-producing activity such as manufacturing, processing.

Chunauti

In August 2020, Chunauti (Challenge Hunt Under NGIS for Advanced Uninhibited Technology Intervention) is a start-up challenge initiative under the Next Generation Incubation Scheme (NGIS). It invite Startups to develop solutions for problems amid pandemic. The government budget of over INR 95 crores over a period of 3 years for this program. This programme aims to identify around 300 startups working in areas. The startups selected through the challenge would be provided incubation support, mentorship, all relevant facilities, as well as funding support such as cloud credits from leading third-party service providers, seed funding of up to INR 25 lakhs, as well as INR 10,000 in monthly monetary support to each intern (startup in pre-incubation) for a period of six months. India seems to be going towards the era of entrepreneurship, as India ranks 63 in ease of doing business around the world.

Digital India Bhashini

On July 4, 2022, PM Narendra Modi announced along with a slew of other initiatives like Digital India GENESIS and India stack, Indian startups the Digital India Bhashini scheme aims to help the Indians with easy access to the internet and digital services in vernacular languages.

Digital India Genesis

Digital India GENESIS is another scheme that Modi launched on July 4, 2022. In this scheme implies "Gen-next Support for Innovative Startups". It is a government of India scheme that is launched to empower the national deep-tech startups.The total budget of this programme and the Digital India Bhashini is disclosed, which is around ₹750 crore.

India Water Pitch-Pilot-Scale Startup Challenge

The scheme launched on March 12, 2022, India will get a grant of ₹20 lakhs in three tranches of ₹ 5 lakh, ₹7 lakh and ₹8 lakh respectively on fulfilling certain conditions/ milestones of work as per their project proposal. While announcing the water startup initiative at an event in New Delhi, under Atal Mission for Rejuvenation and Urban Transformation (AMRUT) 2.0, Minister Puri declared that this initiative the water sector startups and lead them towards growth "through innovation and design that will drive sustainable economic growth and generate employment opportunities."

Ministry of Skill Development and Entrepreneurship

In 2014, the PM decided to dedicate an entire ministry to building this sector as he felt that skill development required greater push from the government's side.

Furthermore, the idea is to reach 500 million people by the year 2022 through gap-funding and skill development initiatives.

These schemes include: (1) The National Apprenticeship Promotion Scheme (NAPS), and (2)MUDRA loans under the Pradhan Mantri MUDRA Yojana (PMMY),

Atal Innovation Mission

In the budget session of 2016, the Indian government announced the Atal Innovation Mission (AIM) is a flagship initiative set up by the NITI Aayog; The government allocated AIM around INR 150 crores in the year 2015.

AIM's goals include building 10,000 Atal Tinkering Labs (ATLs) and 101 Atal Incubation Centers (AICs), as well as sponsoring 200 businesses through the Atal New India Challenges until March 2023.

Academic Applicants are (a) Universities and colleges affiliated with UGC and AICTE, (b) Technical Diploma Colleges. Non-Academic Applicants are Organisations promoting science and technology.

eBiz Portal

It was founded in January 2013. eBiz was the first electronic government-to-business(G2B) portal, The main purpose was to transform and develop a conducive business environment in the country. It was developed by Infosys in a public-private partnership model. It had launched

29+ services in over 5 states of India. As of March 2023, GeM had over 63,000 government buyer organisations, over 6 crore sellers and service providers offering a wide range of products and services. Currently, government departments, ministries, public sector units, state governments, and central armed police forces are allowed to carry out transactions through this portal. Sellers who have registered on an e-Commerce portal to sell items online and have a track record of at least six months.

Dairy Processing and Infrastructure Development Fund (DIDF)

The Government of India announced the creation of the (DIDF) under NABARD in the Union Budget of 2017-18. The total corpus for this fund is INR 8000 crores over a period of 3 years (i.e. 2017-18 to 2019-20). Milk Unions, milk cooperatives, state dairy federations, milk-producing companies, and NDDB subsidiaries meeting the eligibility criteria under the project can borrow loans from NABARD. The loan component would be 80% (maximum rate) with the end borrower's contribution at 20% (minimum rate). The interest rate of 6.5% per annum, repayment will be 10 years. The departments that come under the agriculture ministry were allocated a total of ₹1.31 lakh crore in the 2021 Budget, which has been increased to ₹1.32 lakh crore in the Union Budget of 2022. A fund consisting of blended capital that will be raised under the co-investment model will be facilitated via NABARD, which will finance the agritech startups and rural enterprises that are relevant to farming.

Design Clinic Scheme for Design Expertise

It is a scheme declared by the Indian government to support the MSME (micro/small/medium enterprise) manufacturing sector of India. MSMEs should develop design, innovation and a design-centric approach to fuel their startups, it aims to infuse design expertise in them. Under this scheme, the government of India announced to extend around ₹60,000 for attending design seminars and up to ₹3.75 lakhs or 75% of the cost that would be needed in conducting the seminar, where the entrepreneurs and their teams can learn about design theories, interact with design veterans, build a network, and put them into practice.The Designer in this programme must be chosen from a pool of skilled industrial designers who have been authorised for this programme.

Credit Linked Capital Subsidy for Technology Upgradation (CLCSS)

It was launched on 1st October, 2013.It aims to upgrade technology. Under CLCSS the Indian government provides cash for the Indian companies upfront to upgrade/modernize their equipment or techniques. This scheme mainly empowers Micro and Small-Scale Enterprises (MSMEs). In 2023 with this scheme, an eligible MSME can avail of a credit amount ranging from ₹1 lakh to ₹5 crores. Available from both traditional institutional lenders and NBFCs, a business owner can get this MSME loan at a yearly interest rate of 8.50%. Both new and existing enterprises can avail subsidy under this scheme.

Sub-Mission on Agricultural Mechanization (SMAM)

It was launched by the Government of India in 2014-2015, aiming to increase the reach of farm mechanisation, especially to the small and marginal farmers and in other such areas where the availability of farm power is low. According to the recent updates, as per the reports dated April 2, 2022, the Indian government has announced that it would extend the financial assistance of the full cost of the drones (100%) together with the contingent expenditure under SMAM.

Drone Shakti

It was founded in 1, February, 2022. It is designed to motivate startups to introduce diverse applications for drones, including drones-as-a-service (DaaS). The government would encourage the use of drones by farmers to survey the lands and spray insecticides along with digitising land records. It recently flagged off 100 Kisan Drones in diverse parts of the country. The drone manufacturing industry, as estimated by MoCA, is expected to receive an investment of over ₹5,000 crore over the next 3 years, which will include drones and drone components. The annual turnover of the sales of the drone manufacturing industry, which was valued at ₹60 crore during FY 2020-21, will shoot over to ₹900 crore during FY 2023-24.

Stand Up India Scheme

It was launched by the government of India on 5 April 2016 for financing SC/ST and/or women entrepreneurs over the age of 18. According to the scheme, bank loans between 10 lakhs to 1 crore can be borrowed by at least one Scheduled Cast (SC) or Scheduled Tribe (ST) borrower and at least one woman per bank brand to set up a greenfield enterprise. The greenfield enterprise may be based out of manufacturing, services or the trading sector. Achievements (as on March 21, 2023) ₹40,710 crore has been sanctioned under Stand Up India Scheme to 180,636 accounts and 33,152 crore has been sanctioned to women under Stand Up India Scheme.

Loan for Rooftop Solar PV Power Projects

The government of India has decided to set up 40,000 MWp of Grid-Interactive Rooftop Solar PV Plants in the next five years. It will be set up in residential, commercial, industrial, and institutional sectors in the country and shall range from 1 kWp to 500 kWp in terms of capacity. Such rooftop plants are economically viable since they can produce electricity using solar energy at about INR 7 per kWh without any subsidy. The government also provides a subsidy of 15% on these plants to the associations or individual companies, making the scheme even more lucrative.

OBJECTIVES

1. To study the various schemes of centre government for Entrepreneurship development.
2. The second objective of this paper is to study the effectiveness of government schemes.

METHODOLOGY

The present study is based on a review of different study related to the various government schemes that is launched for entrepreneurship development. The data and information collected from secondary sources like journals, periodicals, reports, textbooks and websites. Table and graph are used for analysis purpose. The main focus is based on the financial funding regarding entrepreneurship development.

CRITICAL ANALYSIS

Name of the Scheme	*Agenda/Purpose*	*Funds*	*Successful Rate*	*Year*	*App name*
Aatmanirbhar Bharat App Innovation Challenge	The unsung digital warriors who had the potential to empower Brand India ,aim to provide financial assistance, mentorship, and support to the budding entrepreneurs "Made in India" application	To identify the best Indian App.The winners will be given the following prize money: First Prize 20 Lakhs, Second Prize15 Lakhs Third Prize 10 Lakhs	NR	4th July 2020	Caption plus, Meme Chat, FTC talent
Samridh	The scheme will support to select and accelerate potential IT based startups.	Funding of up to ₹40 lakh	(F) Around 90 percent tech start-ups in India have reported a decline in revenue • About 30-40 percent have suspended operations or will close down because of the Corona virus pandemic • Companies are not investing due to market uncertainty and Start-ups not achieving the milestones	25th August 2021	"Technology Incubation and Development of Entrepreneurs (TIDE 2.0)", "Centres of Excellence (coes)" in emerging technologies, "meity Startup Hub (MSH)", "Next Generation Incubation Scheme (NGIS)"
Startup India Seed Fund	To provide financial assistance to startups for proof of concept, prototype development, product trials, market entry and commercialization.	An outlay of INR 945 crore	NR	16th January 2021	NR
Startup India Initiative	The objective of supporting entrepreneurs, building a robust startup ecosystem and transforming India	945 cr.	There are over 99000+ startups recognized by the government of India as of May 2023	16th January 2016	SISFS, FFS, CGSS
Startup Leadership Program	Government initiatives that offer help for business start-ups.	NR	This startup initiative has educated over 3,900 people in over 14 countries and helped over 2000 startups.	2016	NR
Pradhan Mantri Mudra Yojana	To support small businesses in India	Government allocated INR 10,000 crores	The last five Years, has benefited 24.48 crore loan accounts with a sanction of ` 12.30 lakh crore	April 8, 2015	• Shishu (up to ₹50,000), • Kishore (between ₹50,000 and ₹5 lakh) and • Tarun (₹10 lakh)

Name of the Scheme	*Agenda/Purpose*	*Funds*	*Successful Rate*	*Year*	*App name*
Chunauti	Is an online Challenge under NGIS to find the products &solutions, issues/challenges faced during and post pandemic situation. It is identify and support the solutions which can support Government, industries & public	Budget of over INR 95 crores	India ranks 63 in ease of doing business around the world	August 2020	Edu-Tech, Agri-Tech & Fin-Tech Supply Chain, logistics & Transportation Management 3. Infrastructure & remote monitoring 4. Medical Healthcare
Digital India Bhashini	Easy access to the internet and digital services	NR	BHIM, UPI, DIGI LOCKER AADHAR	July 4, 2022	1000+ pre-trained AI models have been made available on Bhashini platform
Digital India Genesis	To empower the national deep-tech startups	₹750 crore	Total number of technology startups is 27,000 out of which 3,200+ are deep tech startups	July 4, 2022	Technology Incubation and Development of Entrepreneurs (TIDE 2.0) Scheme
India Water Pitch-Pilot-Scale Startup Challenge	This initiative the water sector startups and lead them towards growth	Grant of ₹20 lakhs	222 entries/applications received, 76 startups were shortlisted and selected on the recommendation of Expert Committee.	March 12, 2022	NR
Ministry of Skill Development and Entrepreneurship	Aims to remove the disconnect between demand and supply of skilled manpower, to build the new skills and innovative thinking	NR	The idea is to reach 500 million people by the year 2022	26 May, 2014,	NAPS, PMMY, UDAAN
Atal Innovation Mission	To develop new programmes fostering, create an umbrella structure innovation in different sectors of the economy, partnership with the Railway Board, Ministry of Drinking Water and Sanitation,	INR 150 crores	NR	2016	NR
eBiz Portal	To transform and develop a conducive business environment in the country., Government-to-Business (G2B) services to business entities with transparency, speed, and certaint	NR	GeM had over 63,000 government buyer organisations, over 6 crore sellers and service providers offering a wide range of products and services.	January 2013	NR

Name of the Scheme	*Agenda/Purpose*	*Funds*	*Successful Rate*	*Year*	*App name*
Dairy Processing and Infrastructure Development Fund (DIDF)	Milk Unions, milk cooperatives, state dairy federations, milk-producing companies, and NDDB subsidiaries meeting the eligibility criteria under the project can borrow loans from NABARD	INR 8000 crores	With the scheme investment, 95,00,000 farmers in about 50,000 villages would be benefitted.	2017-18	Cattle feed/ feed supplement plants Milk transportation system (Refer van/ insulated tankers etc) Marketing infrastructure (including e-market system
Credit Linked Capital Subsidy for Technology Upgradation	Government provides cash for the Indian companies upfront to upgrade/modernize their equipment or techniques	₹40 lakh to ₹1 crore (L) 15%, Nodal Agencies reaches ₹600 crore	sole Proprietorships, Partnerships, Cooperative societies, Private and Public limited companies, Women entrepreneurs	1st Oct, 2013	NR
Sub-Mission on Agricultural Mechanization (SMAM)	Aiming to increase the reach of farm mechanisation, especially to the small and marginal farmers	Government of India funds 60% of the cost and the states' share is 40%	NR	2014-2015	Self Help Groups (SHGs)
Drone Shakti	To motivate startups to introduce diverse applications for drones, including drones-as-a-service (DaaS).	₹120 crore	NR	1 Feb. 2022	NR
Stand Up India Scheme	Financing SC/ST and/or women entrepreneurs over the age of 18	Bank loans between 10 lakhs to 1 crore	Achievements (as on March 21, 2023) ₹40,710 crore has been sanctioned under Stand Up India Scheme to 180,636 accounts and 33,152 crore has been sanctioned to women under Stand Up India Scheme	5 April 2016	NR
LOAN FOR ROOFTOP SOLAR PV POWER PROJECTS	Set up in residential, commercial, industrial, and institutional sectors in the country and shall range from 1 kWp to 500 kWp in terms of capacity	Every bank have different schemes	Subsidy amount: ₹78000/- To be claimed by borrower (loan account number to be provided for credit of subsidy amount in loan account)	NR	NR

As per the table Govt has launched various scheme to focus on entrepreneurship development. Maximum schemes focus for financial assistance through various activities like, app; competition prizes and subsidies. The schemes provide ₹12 Cr. 60 lakh approximately for the growth of entrepreneurship. The schemes has reached to more than 500 million people and they are benefited with the schemes also. The schemes has covered the almost all the industrial sector so that the 360 degree growth can be achieved. So the success rate is quite satisfactory of the govt schemes.

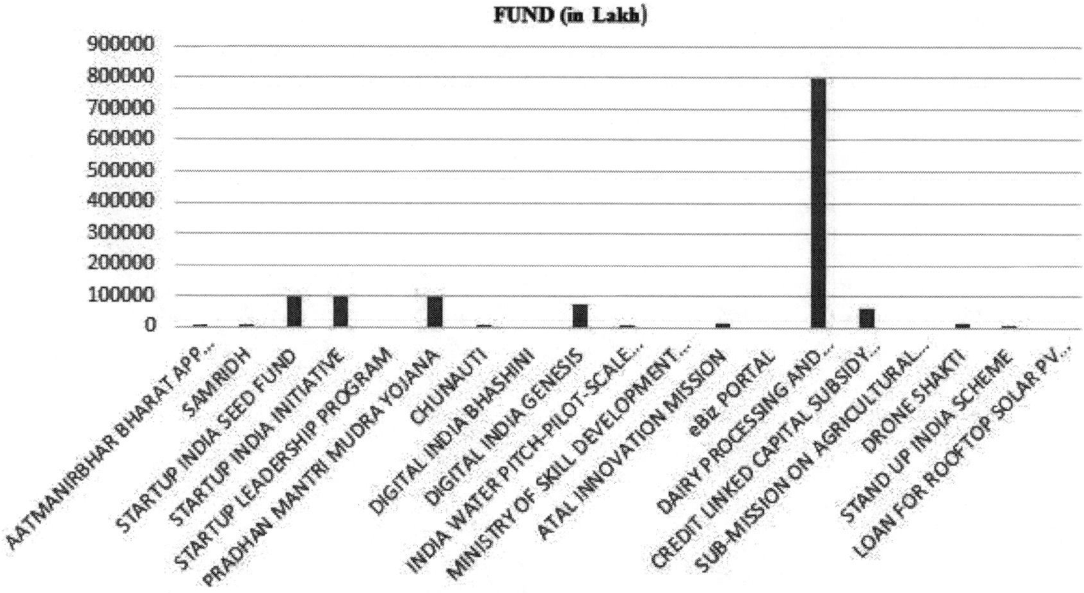

Figure 1

In critical analysis, graphs often serve as valuable tools for representing various parameters such as high, low, and average values. Graphs provide a visual representation of data, making it easier to identify trends, patterns, and relationships among different variables. By plotting data points on a graph, analysts can quickly discern the range of values, spot outliers, and understand the distribution of data. In this graphical representation "Dairy Processing And Infrastructure Development Fund" enhances and explore the programme and many people got benefit from this scheme. Some of the schemes doesn't explore the scheme of many reasons.

CONCLUSION

The conclusion of a critical study on the role of government schemes for entrepreneurship development would likely summarize the key findings and insights gathered throughout the research. This paper focus on various government scheme to the development of Entrepreneurship efficiency. The entire schemes are most beneficial for the Micro, Small and Medium Enterprise (MSME). Government schemes play a crucial role in providing essential support structures, such as access to finance, mentorship, training, and networking opportunities, which are vital for nurturing aspiring entrepreneurs and sustaining existing ventures. Our study also highlights several challenges and areas for improvement like the proliferation of schemes, bureaucratic inefficiencies, lack of awareness, and inadequate monitoring and evaluation mechanisms often hinder their effectiveness. Government schemes represent a vital tool for fostering entrepreneurship and driving economic growth.

REFERENCES

Kale Santosh Subhash, & Rajiv Babanrao Mohite (2019). A Study of Government role in Entrepreneurship Development with Special Reference to Marathada Region. Ajanta Prakashan, Vol. VIII, Issue I, pp. 168-172.

Bhagyalaxmi M., & Dr Ishwara P. (2014). Dual Responsibility of Rural Women Entrepreneurs – Issues and Challenges, *GJRA Global Journal For Research Analysis*, Vol. 3, Issue 7, ISSN No 2277-8160.

Elizabeth J. Gatewood et al. (2009). Diana: A Symbol of Women Entrepreneurs' Hunt for Knowledge, Money, and the Rewards of Entrepreneurship, Springer Small Business Economics February, Volume 32, Issue 2, pp. 129–144.

T. Swetha & Dr. K. Venugopal Rao (2013). "Entrepreneurship in India", *International Journal of Social Science & Interdisciplinary Research*, Vol. 2(7), July 2013.

Amit (2014). "Entrepreneurship development in India: The role of economic growth, foreign investment and financial development", *AIMA Journal of Management & Research*, August 2014, Volume 8, Issue 3/4, ISSN 0974–497.

Dzafic Z., & Babajic A. (2016). "The role of the government in entrepreneurship development: evidence from Bosnia and Herzegovina", *Economic Review – Journal of Economics and Business*, Vol. XIV, Issue 1, May 2016.

Jasmer Singh Saini (2003), Entrepreneurship Development Programmes and Practices, Deep & Deep Publications Pvt Ltd, New Delhi.

https://msme.gov.in/

https://www.sidbi.in/About_SIDBI.php

https://www.tutorialspoint.com/entrepreneurship_development/entrepreneurship_development_introduction.htm

https://www.researchgate.net/publication/371957855_Role_of_Government_Policy_in_Entrepreneurship_Development

CHAPTER 30

In-Service Teachers' Training and NEP 2020

ANAND S.S. KANDERI

Dean, Student Welfare Committee, Himalayan Institute of Technology, Dehradun, India

ABSTRACT

New Education Policy 2020 has been implemented in the education system of India recently. The policy has been hailed one of the best policies so far made in India by many renowned educationists. Many rounds of seminars and key discussion have been held on different corners of the country. Many thoughts have been put on the quality of the educators who will be the key players of the policy at grass root levels. There is substantial pressure from every corner to make its implementation quicker and effective. The teachers, who are the flag bearers in implementation of the policy, are supposed to be well qualified, updated and equipped with latest technology in their hands. Their efficiency should be capable to bring drastic and comprehensive changes in the entire system of education in India, for which an immediate need of trainings to educators is felt to make the implementation of NEP 2020 a must exercise to get positive outcomes.

5+3+3+4 education system is going to change the entire system of teaching in the classrooms. Teachers would be taking care of the child's over all development and progress. The child is going to be trained enough to become "ATMANIRBHAR" adult after completing the 15 years of Schooling. In a nutshell, the teachers of all forms must get immediate in-service training to implement the NEP 2020.

INTRODUCTION

New Education Policy 2020 has been strongly advocated by the educationists, government and stakeholders in recent past. There is substantial pressure from every corner to make its implementation quicker and effective. Since its origin many seminars, debates, capacity building programme and conferences are the regular exercises undertaken by different education agencies on NEP 2020. In-Service teachers' training based on NEP's theory will certainly help the education sector as a whole to get the maximum out- put of NEP's implementation. Skills, knowledge and experience of the teachers is desirable to get maximum essence of reform only when teachers are well trained. The effectiveness of the implementation solely depends on the execution part of the policy inside the class rooms and enhancement of overall standard of education. The teachers, who are the primary source of implementation of the policy, need to be well aware and equipped to carry forward the light of new innovations, for which, there should be immediate exercises to train the teachers before they apply these new methodologies in their teachings.

Higher Education and Teachers In-service Trainings

Higher education system of India has hardly any large institutions for faculties in-service trainings, just like NCERT and C.B.S.E. which provide trainings for in-service teachers of pre-primary to senior secondary teachers across the country. The NEP 2020 is one of the series documents (1968 and 1986-92) which made roads for the country to 21st century. However, numbers of seminars and conferences were arranged to make strategic roadmap to implement the objectives of higher education for 21st century.

Indian higher education has Higher Education Commission of India (HECI) which is branched to National Higher Education Regulatory Council (NHERC), National Accreditation Council (NAC), Higher Education Grants Council (HEGC) and General Education Council (GEC) and these all bodies are the responsible to take forward the objectives of the NEP 2020. But, so far, no platform has been set by these bodies to train the working teachers of universities and colleges to implement NEP 2020.

As on today, there are 405 State Public Universities and 358 State Private universities, along with over 40,000 colleges in India. Tens and thousands of faculties working in these institutions have to take some kind of trainings on NEP 2020, which is a hurricane task just before starting of the new session. It would be felt necessity to the faculties to enter in the class with some prior knowledge of implementation and teaching methodology of the NEP 2020.

Private and Government universities are arranging seminars, research papers and FDPs in limitations, however, so far they have no set guidelines like compulsory teachers' in-service training in a calendar year like C.B.S.E. schools under go in India, where every teacher shall undergo 5 days of in service trainings in a single calendar year. NIEPA (The National Institute of Educational Planning and Administration, a Deemed to be University) does some capacity building and research in planning and management of education not only in India but also in South Asia. However, it is not enough such as country like India. However, NEP 2020 document says as 15.10 (NEP 2020) "In-service continuous professional development for college and university teachers will continue through the existing institutional arrangements and ongoing initiatives; these will be strengthened and substantially expended to meet the needs of enriched teaching learning processes for quality education. The use of technology platform such as SWAYAM /DIKSHA for online training of teachers will be encouraged, so that standardized training programme can be administered to large number of teachers within a short span of time."

School level Teachers in-service Trainings

The NCERT, is the main body in India, which provides trainings for educators up to Senior secondary teachers. SCERT and DIET are the other training bodies which are providing trainings to teachers in State Education Boards and District Level Education Educators. So far these bodies are working superbly and comprehensively for all Government teachers. However, the private school teachers are not getting any kind of trainings from these institutions. In school level, NISTHA, (National Initiative for School Heads' and Teachers' Holistic advancement) has done a commendable job in past few years where NISTHA Elementary face to face, Elementary online, Secondary online, FLN for NIPUN Bharat-online, NISTHA ECCE-online have provided trainings and going to provide massive number of teachers across India. NCERT has been nominated as the National Level Coordinator to develop massive Open Online Courses (MOOCs) for school education (Class IX to XII) by Ministry of Education, as part of this initiatives till date, 28 courses were developed and are being popularized through "SWAYAM" portal and mobile app. Twelve cycles have been already completed and remaining will be completed by September 2024.

Framework of Teachers Training Education under NEP 2020

Two years of B.Ed. teachers' training course, basic teacher trainings is being replaced by four years of integrated, multi disciplinary programme which would be fully effective before 2030. Trained teachers should be fully trained as per the learning theories of education framed according NEP 2020.

The Foundation Literacy and Numeracy (FLN) mission under NIPUN BHARAT

Foundation learning has been incorporated in government and private schools in recent months. Young learners are included in schools BAL-Vatika -3, pre-KG, LKG and UKG where minimum age limit is young as 3 years. A huge number of trained teachers are required to teach these most significant classes of education immediately, where so far, we have few in numbers who have taken trainings from NCERT's mission training like NISTHA, DIKSHA etc., or other training agencies during recent years.

5+3+3+4 education system is going to change the entire system of teaching in the classrooms. Teachers would be taking care of the child's over all development and progress. The child is going to be trained enough to become "ATMANIRBHAR" adult after completing the 15 years of schooling.

Advantages of Teachers Getting In-Service Training on NEP 2020

A trained teacher under NEP 2020 can be highly effective and useful to carry forward the objectives of NEP 2020 in the real class room by following ways:-

1. Disseminating the core objective of NEP 2020 in the classroom.
2. Change the teaching methods as per requirement of the NEP 2020.
3. Gets familiar of the changed syllabus and methodology.
4. Understands child/ student's psychology and learning instincts.
5. Understands child/ student's areas of interest.
6. Becomes able to know the life skills and job skills of the learners.
7. Learns latest theories of software, computing, artificial intelligence and its uses in curriculum.
8. Provides freedom to the learners so that the learners take decisions as per their potentials and requirement.
9. It increases the numbers of the trained and effective teachers in India.
10. It provides updated knowledge and latest scenario in Education at International Level.
11. It encourages educators to learn number of Indian and foreign languages.
12. Educators can provide career counseling to all the learners as per their capability and areas of interest.
13. It encourages hybrid, blended learning and art integrated learning.
14. Educators become best facilitators inside the classrooms.

Suggestive Subjects of In-Service Teachers Training

After implementation of NEP 2020, all Indian teachers, right from FLN to higher education, should have immediate in-service trainings on following areas:

1. Modern pedagogy of NEP 2020.

2. Integrated learning with technology.
3. Modernized rural agriculture developmental program.
4. Art integrated education.
5. Native and Multilingual Indian and foreign Languages.
6. Sports and Games, yoga, generic sports and traditional rural Indian sports.
7. Social and emotional Learning.
8. Artificial Intelligence and machine learning.
9. Critical thinking and problem solving skills.
10. National values, ethics and value education.
11. Cyber safety and security.
12. Science and Technology.
13. Health and wellness.
14. Sustainable environmental education.
15. Child psychology and learning by doing.
16. Adolescent Education.
17. Life skills.
18. Freedom of learning, allowing students to make frequent mistakes in early education.
19. Generic subjects focusing on gender equality, adult education and peace education.
20. Regular FDPs and in-service trainings on updating implementation of NEP 2020.

CONCLUSION

In a nutshell, it is emphasized that teachers of all categories in India must get some comprehensive training covering all important objectives of NEP 2020 so that the learners of the nation become skilled and independent when they complete their education under the umbrella of NEP 2020. The nation looks forward to see a massive quality shift in Education from Degree oriented to skill oriented through the implementation of NEP 2020.

REFERENCES

Bhatia Rainu, Rinkey and Kaur Jagneet, Reserch Scholars, Department of Education, Etwrnal University, Baru sahib, Sirmour, HP, India. ISBN 13: 9788119907274.

C.B.S.E. Teacher's resource for advance learning outcomes (2019).

Das Tapash and Halder Tarini 2023, Teacher Education for Transforming Society, First Edition , Red'shine Publication.

MHRD- Government of India (2019) – Ministry of Human resource development – New Education Policy 2019 booklet.

NCERT (2009), National Curriculum Framework of Teacher Education, New Delhi.

National education Policy 2020. Ministry of Human Resource Development, Government of India, Published in the Government of India Gazette, 29 July 2020.

Vedanayagam, E. G., 1966. An Analysis of the In-Service Teacher Education Programme for Secondary Schools in India. OREGON STATE UNIVERSITY, Ph.D. Thesis.

CHAPTER 31

Re-imagining Teacher Training: Co-creation Workshops and NEP 2020

Sonali Verma

TGT Natural Science, Government Girls Senior Secondary School, DoE, GNCTD, New Delhi, India

Abstract

Various research and policy documents have highlighted the inadequacies and limited impact of current in-service teacher training scenario in India. NEP 2020 has taken cognizance of this fact and has recommended shifting from annual one-time in-service teacher training to a Continuous Professional Development (CPD) model. NEP 2020 has laid importance to adjust teaching practices that address creative problem-solving, critical thinking and experiential learning rather than rote memorization. To implement these pedagogical approaches and making our teachers successful in imparting 21st century skills including collaboration, etc., it is of utmost importance that our teachers also experience a transformation in the training practices. Traditional teacher training programs have been utilizing passive lecture-based techniques, which seem inappropriate as far as progressive educational paradigm is concerned. This paper puts forward the idea that Co-Creation Workshops present a practical and effective solution in this direction. These workshops facilitate collaborative, hands-on learning experiences for teachers, closely aligning with the active learning strategies they are expected to adopt and teach. By engaging in co-creation workshops, teachers can directly experience and practice the methods they will employ in their classrooms, so that the principles of experiential learning and collaborative problem-solving are internalized by them. Co-creation workshops provides opportunity to teachers to be an active participant in the design of the training module as per their needs and demands of their professional development journey, hence inducing a sense of ownership in their own learning. Co-creation workshops also ensures that teachers are better equipped to foster a dynamic and interactive learning environment for their students.

Keywords: *Co-creation Workshops, Teacher Training, Professional Development*

INTRODUCTION

Teacher as a Professional

Teaching is a profession. It is an occupation where individuals are trained extensively in techniques and methodologies before being imparted duties as a teacher. Society places a lot of trust in teachers when it hands over it younger generation in the care of these individuals. With the changing times, there has arisen an imperative need in adjusting the profession to the newly created aspirations and demands that manifest in the societies. There has been a tremendous advancement in the available technology that has revolutionized every aspect of human life. Perceptions have been changing and the traditional methods do not seem fit in the drastically changing education system. New frontiers are being explored and successful models from other fields are now finding their place in enhancing the quality of education. Teachers too need to do significant additions in their existing knowledge base in order to suffice these demands of the society so that they can contribute in shaping the individuals that are better adapted to the changing world scenario. This necessitates transforming the in-service teacher education model by proving teachers with sufficient opportunities for their continuous professional development which make them better suited to 21st century. The importance for continuous professional development of teachers have been given due weightage and there have been policies and recommendations that have put this aspect of teacher development at the forefront. Different Commissions in India too have emphasised the importance of continuous professional development for teachers. A few examples from various commissions are mentioned:

- Secondary Education Commission (1952–53) under the chairmanship of Dr. A. Lakshman Swami Mudaliar advocated: "However excellent the programme of teacher-training may be, it does not by itself produce an excellent teacher. It can only engender the knowledge skills and attitudes which will enable the teacher to begin his task with a reasonable degree of confidence and with the minimum amount of experience. Increased efficiency will come through experience critically analyzed and through individual and group efforts at improvement."
- The National Policy of Education (NPE 1986/92) mentioned categorically that "teacher education is a continuous process, and its pre-service and in-service components are inseparable."
- Clarifying the role of in-service teacher education as professional development activity, the Acharya Ramamurthi Review Committee (1990) explicitly advocated that "in-service and refresher courses should be related to the specific needs of the teachers. In-service education should take due care of the future needs of teacher growth; evaluation and follow up should be part of the scheme."
- Report of the high-powered Commission on Teacher Education constituted by Hon'ble Supreme Court of India (2012), titled "Vision of Teacher Education in India: Quality and Regulatory Perspective", commonly known as Justice Verma Commission, suggested to set up a model of INSET policy which should focus on continuing professional development opportunities for teachers.

CURRENT STATE OF TEACHER TRAINING AND CONTINUOUS PROFESSIONAL DEVELOPMENT

Currently, a cascading model of teacher training and professional development is being utilized for training as far as the teachers at the Delhi Government schools are concerned. It is a top-down approach where the agenda for teacher education is devised and developed by the department responsible for carrying out trainings. There is a flow of information from "Experts" to Master Trainers which then conduct training via workshops, seminars etc for the teachers. These trainings usually focus on imparting content-specific expertise, pedagogical skills, and training teachers on the initiatives that have been implemented by the Directorate of Education like Happiness Curriculum, Entrepreneurship Mindset Curriculum, Deshbhakti Curriculum, etc. These trainings have pre-decided agenda where one-way transmission of information takes place.

Though being quite helpful as far as dissemination of information and skills is required for a large teacher community, as is the case in the government schools of Delhi, the cascade model faces significant criticism. Its primary disadvantage is the distortion of intended message during the training. As there are many levels through which the information passes, there are chances of miscommunication and differing interpretations. It has also been witnessed that the training plans and guidelines are not followed in their true spirit, rather varied strategies and often new elements are introduced. This makes the whole exercise of imparting training futile.

Another major issue is the gap between central and local levels. McDevitt likens it to being "too far from the source to get soaked." Also, the one-way, top-down structure of the cascade model is quite rigid in itself rendering it incompetent to meet grassroots needs. Additionally, it is also felt that it is designed and implemented by those who lack primary teaching experience. This widens the gap between levels.

NEP 2020 on CPD

NEP 2020 recognizes continuous professional development of teachers as a crucial factor in improving the quality of education in India. NEP 2020 makes fifty (50) hours of continuous professional development mandatory every year for all the teachers. They can improve their skills and knowledge base through workshops, seminars, online courses, etc. The policy also puts emphasis on establishing professional learning communities to strengthen collaboration and cooperation among teachers where they can share their best practices and also discuss their challenges. These communities can help them to learn from each other and to reach solutions of the problems faced. NEP 2020 also recognizes the importance of mentoring and peer learning as feasible approaches towards continuous professional development of teachers.

CO-CREATION AS A MODEL FOR TEACHERS' CPD

Co-creation can pave the way for a more dynamic and vibrant alternative for continuous professional development of teachers. Co-creation model of teacher professional development involves active engagement and offers opportunities for contextualized learning. It also fosters an environment and culture of collaboration and cooperation. In this fast-changing landscape of education, it is important that the ground-level practitioners like teachers are imparted chance to decide and develop agenda of their professional development journey depending upon their needs. Embracing co-creation as a model for professional development will empower teachers and will ensure that they are better equipped to meet the diverse needs of their students. In our quest for excellence in education, co-creation workshops need to be at the forefront of our CPD strategies, driving meaningful and sustained improvements in teaching and learning.

What is Co-creation

C K Prahlad and Venkat Ramaswamy have introduced the term co-creation for product and service design process. According to them, co-creation is defined as a business that gives due importance in creating value through interaction between companies and customers. This approach was in contrast to the traditional business models where companies created value and customers consumed it. Co-creation involves customers' active contribution to the business value creation process for outcomes that are symbiotic for both customers and the business. This approach has been successfully transferred to the educational landscape as well. Co-creation model has been utilized for teaching-learning and curriculum development as well. Here major stakeholders, i.e. administrators, teachers, students, etc., have contributed to the development of learning experiences which has led to enhanced ownership and better learning outcomes.

This paper proposes utilization of this methodology of co-creation in designing teacher workshops for more collaborative professional development experience. It is a decentralized approach that promotes:

- Teacher agency
- Teacher autonomy
- Teacher empowerment.

ORGANIZING AND FACILITATING A CO-CREATION WORKSHOP

1. **Preparation:** Need assessment can be carried out by conducting online surveys. Broad themes emerging from the survey can be chalked out and at the initial stage, the facilitator can provide a platform for discussions and safety and confidence to the teachers to present their viewpoint and devise an agenda for the workshops. These emerging themes may vary from subject specific skills to development of digital resources to make teaching learning process more effective. The facilitator should ensure equitable and fair participation while developing the agenda.
2. **Organization:** The workshop can be designed according to the decided agenda. Necessary skills and resource creation can then take place. Participating teachers are the main actors where they share their experiences. The facilitator helps them to brainstorm on the identified challenges and helps them to collaboratively arriving at plausible solutions. Resources can be shared or demonstrated. Insights can be built on new pedagogical approaches by providing ample hands-on experience during these workshops as well. These workshops can also provide a platform for development of repositories and banks of digital and manual resources.
3. **Implementation and Iteration:** The learnings hence gathered in the previous stage are now to be implemented in the classroom setting. The teachers have to plan it according to their context and apply in their classrooms. The necessary resources should be identified while planning and appropriately used. For clear understanding and acquiring excellence in implementation of the practice, an iterative process of planning, action, observation, and reflection needs to be promoted by the teacher. The teachers can carry out formative assessments, low stake quizzes or any other measures to evaluate the effectiveness of the practice. Based on the evaluation, the teacher can make the necessary modification and re-adjustments. The teachers can document their practice.
4. **Sharing and Sustenance:** The challenges and success stories can be shared among teachers through digital or physical platforms. A learning community can thus be established. WhatsApp groups, Insta reels, blogs and vlogs or any other digital platform can help in realizing this objective. A thriving community can be helpful in sustaining the motivation among teachers and acts as a necessary fuel by providing impetus to embark on a journey

of professional development and self-fulfilment. Working in isolation can be a demotivating factor which can sometimes discourage teachers.

ADVANTAGES OF CO-CREATION WORKSHOPS

- **Collaborative Learning:** Various studies and researches have pointed towards the benefits and advantages of collaborative learning. It is helpful in generating critical thinking skills and promotes oral communication, self-management, responsibility and self-esteem. It helps in providing a safe environment where participants can discuss their misunderstandings and doubts and bring clarity to their thought process.
- **Teacher Empowerment:** The teachers are involved in deciding and devising the agenda of the workshops/training modules which promote teacher empowerment and decentralization. The agendas have been crafted collaboratively based on the specific needs of the participating teachers which stirs a feeling of responsibility and autonomy among teachers. This renewed self-esteem through empowerment can go a long way in their professional development and also subsequent translation into better academic achievements by their students.
- **Teacher Leadership:** The model of co-creation professional development offers ample opportunities for teachers to establish themselves in the role of leaders. A leader is a responsible person who knows how to take initiative and lead others (students) towards desired outcomes. Various studies in this field have recognized the linkage between teacher leadership and better student achievement. Teacher leader can act as a catalyst towards growth, improvement and development of his/her students.
- **Teacher Agency:** Agency is a dynamic competence of individuals to act independently and to make their own actions and decisions (Barker 2005). Teacher agency can be established when they act purposefully towards their own professional development and contribute their learnings constructively towards the betterment of quality of education. Co-creation model of professional development by sensitizing teachers towards their professional identity and obligations towards society contributes actively towards development of teacher agency. Teacher agency also plays a crucial role in construction of knowledge.
- **Teacher as Researcher:** Co-creation model of professional development extensively promotes iterative process of planning, action, observation and reflection which are components of action research. Following this practice, teachers engage in research and document the steps taken by them while implementing new approach or strategy thereby creating a repository of resources for the next generation to learn from. Research in their classroom setting enables teachers to take informed decisions that are data driven. This leads to effective teaching learning practice.

REFLECTIONS

Through focus group discussions which were carried out with the teachers, it emerged that teachers who have been part of co-creation workshops felt a sense of ownership in their own learning process. They felt a sense of responsibility and agency. According to a few of them, co-creation workshops provide a safe space where they can discuss their challenges and doubts with their peers without any hesitation or feeling of inferiority. They feel that the space of co-creation workshop provides tailored content that is suited to their needs, and they enjoy the opportunities to learn from their peers. Also, when they listen to the success stories of their peers, they obtain insights as to how to use those learning experiences in their classroom contexts. Co-creation workshop has also helped them in creating various resources including unit plans, digital presentations and have helped them in improving their skills as to how to handle digital devices available at their school level. Most of them voiced that they always look forward to such co-creation workshops and feel emphatic and enthusiastic towards their participation.

CONCLUSION

Co-creation workshops emerges as a unique alternative within the framework of teacher training and their continuous professional development as envisioned by the National Education Policy (NEP) 2020. Also, in the scenario of changing landscape of global education, co-creation model provides opportunities to teachers to first practice 21st century skills and modern pedagogical approaches themselves so that they can effectively transfer them to their students. It reduces passivity and generates active participation thereby promoting and fostering a sense of teacher leadership, teacher autonomy and teacher agency. Co-creation is a modern, equitable and inclusive model of teacher professional development. It provides experiential learning and holistic learning environment which resonates with the spirit of ownership and commitment among teachers towards their own professional journey. It is a dynamic approach rather than being static. The model is adaptable and scalable. It can be adjusted to meet the diverse needs at different regional levels. The institutions entrusted with the development of in-service teacher training modules can opt for the decentralized alternative of co-creation model which treats its participants not as passive recipient of information, rather envisioning them as co-constructors of knowledge.

REFERENCES

Abeysena, H., Philips, R., & Poppit, G. (2016). The Cascade Model in Action. English Language Teacher Research Partnerships. *A collection of research papers from the Sri Lankan context*, 79.

C. Laltlansangi, Cherthangpuii L. Nunchanga, Reflection of NEP 2020 on Professional Development of Teachers. Department of Education, Mizoram University, Tanhril, Mizoram India Institute of Advanced Studies in Education Aizawl, Mizoram, India.

Dimova, Y., & Kamarska, K. (2015). Rediscovering John Dewey's model of learning through reflective inquiry. *Problems of Education in the 21st Century*, 63, 29.

Hayes, D. (2000). Cascade training and teachers' professional development. *ELT Journal*, 54(2), 135-145.

Jeffrey Hsu, Li-Chun Lin, Mel Stern. Curriculum co-creation: Knowledge Co-creation in an educational context.

Louis Langdon Warren. The Importance of Teacher Leadership Skills in the Classroom. *Education Journal*, Vol. 10, No. 1, 2021, pp. 8-15. doi: 10.11648/j.edu.20211001.12.

McDevitt (1998). How effective is the cascade as a method for disseminating ideas? A Case Study in Botswana. *International Journal of Educational Development*, Vol 18, Issue 3.

MHRD (2020). National Education Policy 2020. Retrieved from https://www.mhrd.gov.in/sites/upload_files/mhrd/files/NEP_Final_English_0.pdf

Takako SUZUKI:The Effectiveness of the Cascade Model for In-service Teacher Training in Nepal. Graduate School of International Cooperation Studies, Kobe University 2-1 Rokko-dai, Nada-ku, Kobe, 657-8501 Japan.

Tshewang Rabgay, Gillian Kidman. Multiple iterations and messiness in the implementation of action research by Bhutanese secondary science teachers.

https://link.springer.com/article/10.1007/s44217-023-00077-4.

https://teaching.cornell.edu/teaching-resources/active-collaborative-learning/collaborative-learning.

CHAPTER 32

The Strategic Execution of the NEP

Rajesh Singh Negi

Training and Placement Officer, Himalayan Institute of Technology, Dehradun, India

Abstract

The inception of this paper/research stems from the fact that, the successful implementation of a comprehensive policy like the New Education Policy (NEP) depends on effectively understanding the target audience, i.e. students, since students possess diverse personalities. Therefore, concentrated efforts are required to motivate them, to embrace the journey outlined by the NEP. The responsibility for this endeavor lies with the guiding force of education, i.e. gurus (teachers) and the temple of education, i.e. Institute. In today's evolving world specifically "Generation Z", the concepts like independence, choice, fulfillment, aspirations, and desires have advanced, presenting new challenges to the role/power of education to ensure E2 factors, i.e. employment and empowerment. Furthermore, human desire resides within, which ignites/enhances due to biological changes and the kind of exposure one witnesses. As widely recognized, the boundless scope of the internet and social media has rendered desire elusive and devoid of practicality. Besides due to rapid changes in the world driven by frequent innovations and technological advancements, the widening gap between practical experience and theoretical understanding has heightened the challenges in implementing educational policy.

Role of "Atmanirbhar Bharat" Fostering comprehensive development beginning with the Rigveda era, our ancient educational system gradually evolved, prioritizing the comprehensive development of individuals by nurturing both their inner and outer selves. This system emphasized moral, physical, spiritual, and intellectual facets of life. Undoubtedly, fostering an Atmanirbhar Bharat through the implementation of the new education policy will serve as a guiding light for today's generation.

Therefore, like in Business schools before, setting up a business model there is a greater need to understand the Target audience i.e. youth who are participating in the carnival of education and equipping flag bearers of education i.e. teachers (gurus) to ensure effective execution of the new education policy i.e. NEP -2020.

Keywords: *Generation Z, Target Audience, Flag Bearer of Education, i.e. Teachers (Guru), Atmanirbhar Bharat*

INTRODUCTION

The objective of this research ensure comprehensive readiness and effective preparation by leveraging the ancient Indian educational methodology, which prioritizes comprehensive development for the successful implementation of the NEP. Predominantly understanding the ancient education system of India where students resided with their guru and learned practical knowledge applicable to real-life problems. A strong emotional bond between the guru and shishya was essential before engaging in the teaching-learning process. Education was not about memorizing facts and figures for exams but about understanding and connecting with nature and life.

METHODOLOGY

Regarding the methodology adopted, an in-depth study has been conducted on the ancient Indian study patterns, particularly emphasizing the creation of a bond between the guru and shishya before commencing the curriculum. Additionally, other papers have been reviewed to understand the modern educational requirements, keeping in mind the continuously evolving industry demands. In a comparative study of factors such as ancient Indian education, modern educational requirements, and the target audience (students participating in the educational process), the following points must be considered before commencing the curriculum as directed by NEP 2020 for effective implementation:

- Understanding students within the context of Indian continental
- How to ensure an environment that enhances youth's active participation in the process of education
- How to balance between particle exposure and theoretical understanding
- How to equip flag bearer of education i.e. teachers(guru) to ensure success full implementation
- How to ensure Physiological / Psychosocial aspects of youth.
- Ensure educational institute infrastructure to support skill-based education.
- How to ensure active industry participation.

Understanding Students within the Context of Indian Continental

Our social landscape is distinct from the rest of the world, making it evident that policies successful elsewhere may not be suitable here. This is particularly true for our diverse student population.

Considering the target audience from an Indian continental perspective, the focus of this paper presentation is "The Strategic Execution of the NEP," aiming to achieve the E2 factor: Employment and Empowerment. Our students come from varied backgrounds, especially in terms of family background and income status. Therefore, imparting knowledge must begin with a planned orientation session. Before the commencement of the formal curriculum, it is essential to understand the target audience by assessing family background, socioeconomic status, personality traits, and physiological and psychological aspects. The proposed orientation plan with the following points may be considered for the strategic execution of NEP.

- Background understanding
- Socioeconomic status
- Physiological and psychological understanding
- Emotional assessment
- SWOT analysis
- Personality assessment

Strategy for Youth's Active Participation in the Process of Education

A comprehensive understanding of youth participation and its implementation is essential for the successful execution of NEP. Additionally, it's critical to establish a set of measurable indicators to supplement the objectives and benchmarks created for assessing progress. Presently Ensuring youth's active participation in the academic journey is challenging, particularly in the context of its volatility, uncertainty, complexity, and ambiguity (VUCA). To engage young people effectively before the start of the curriculum. it's essential to plan interactive methods keeping the Student-Centered Approach. Certain methods can be incorporated to ensure youth active participation.

- Interactive Teaching Methods
- Incorporate Technology
- Student-Centered Approaches
- Peer Learning and Mentoring
- Extracurricular Activities

Strategy for Making a Balance between Particle Exposure and Theoretical Understanding

The applicability of education is two-fold, i.e. empowerment and employment. When we talk about empowerment it exactly means the enhancement of skills of the individual. Today's world is rapidly changing what is new for today has become old tomorrow. there is rapid advancement and innovation taking place therefore is nearly impossible to change the curriculum with respect to changes in the industry environment. Now the question arises of how frequently the amalgam between industry platform and institution should take place and how to ensure that minimizing the widening gap between particle exposure and theoretical understanding. The following points can be considered.

- Hands-on Projects
- Internships and Work Experience
- Case Studies
- Field Trips and Site Visits
- Problem-based learning
- Simulation Exercises
- Mentorship Programs
- Reflective Practice

Strategy for Equipping the Flag Bearer of Education, i.e. Teachers (Guru)

At the core of this ancient Indian education system was a profound bond, where the Guru was not merely an instructor but a respected mentor and guide therefore establishing this bond is very essential since we are in the VUCA world, which stands for volatility, uncertainty, complexity, and ambiguity. Notably A holistic approach to education extends beyond academic instruction, aiming to nurture each student's cognitive, emotional, social, physical, and spiritual development. Hence it is quite imperative that for the successful implementation of NEP, the flag bearer of education i.e. teachers(gurus) need to gain those skills that propel students' active participation in the academic journey. following are some of the methods that need to be incorporated where the guru creates that bond less like a teacher but more like a respected mentor and guide

- Cultural Sensitivity and Inclusivity
- Supportive Work Environment
- Continuous Professional Development

- Integration of Technology
- Incentivize Further Education
- Personalized Professional Development Plans

Strategy for Understanding Physiological/Psychosocial Aspects of Students

The physiological and psychosocial aspects of students are essential for promoting their holistic development, supporting their well-being, and creating inclusive and supportive learning environments. Awareness of physiological and psychosocial diversity among youth, including factors such as culture, gender identity, and socioeconomic status, can inform efforts to create inclusive learning environments. Insight into the psychosocial influences on youth behavior can help educators implement preventive measures against risky behaviors such as substance abuse, violence, and unsafe sexual practices. Educators can provide education, guidance, and support to empower youth to make healthy choices. Hence teachers and staff need to be trained to recognize and respond to various physiological and psychosocial aspects of students. The following methods can be incorporated to understand the Physiological / Psychosocial aspects of students:

- **Workshops and Seminars:** Organize regular workshops and seminars on child development, mental health, and physical health. Topics can include common psychological issues (e.g., anxiety, depression), learning disabilities, and physical health concerns (e.g., vision, and hearing problems
- **Certification Programs:** Encourage teachers to pursue certification programs in special education, mental health first aid, or human psychology.
- **Scenario-Based Training:** Assist teachers in practicing the recognition and response to various physiological and psychosocial issues, role-playing and case studies can be highly effective. Develop realistic scenarios that teachers might encounter in the classroom and provide guidance on appropriate responses

Strategy for Educational Institute Infrastructure to Support Skill-based Education

Despite initiatives, viz. The Skill India Mission, inaugurated in 2015, sets its sights on training over 400 million individuals across diverse skill sets by 2022. Within this initiative, several schemes such as the Pradhan Mantri Kaushal Vikas Yojana (PMKVY), National Apprenticeship Promotion Scheme (NAPS), and the Skill Loan Scheme have been put into action however challenges persist in areas such as insufficient infrastructure, and discrepancies between the skills taught and industry demands. It is imperative to tackle these hurdles to ensure the successful execution of skill-based education initiatives.

CONCLUSION

In summation, the successful implementation of the NEP hinges on two key factors: the educators, or gurus, and the students, the target audience. The effectiveness and smooth interaction between these two elements will determine how the NEP progresses, potentially establishing it as a model for the rest of the world

REFERENCES

Ajay Kurien and Sandeep B Chandramana 2020. Impact of new education policy – 2020 on higher education, *research gate*, 02-10

Urmila Yadav,2018. A Comparative Study of Ancient & Present Education Systems, *Zibeline International*,03

Ministry of Human Resource Development, Government of India, New Education Policy – 2020,

Author Index